Eleanor O´Grady

Aids to correct and effective elocution

With selected readings and recitations for practice

Eleanor O´Grady

Aids to correct and effective elocution
With selected readings and recitations for practice

ISBN/EAN: 9783741177910

Manufactured in Europe, USA, Canada, Australia, Japa

Cover: Foto ©berggeist007 / pixelio.de

Manufactured and distributed by brebook publishing software
(www.brebook.com)

Eleanor O´Grady

Aids to correct and effective elocution

AIDS TO

CORRECT AND EFFECTIVE

ELOCUTION

WITH

Selected Readings and Recitations

FOR PRACTICE.

BY

ELEANOR O'GRADY,

Author of "Select Recitations," etc.

———————

NEW YORK, CINCINNATI, CHICAGO:

BENZIGER BROTHERS,

Printers to the Holy Apostolic See.

PREFACE.

THE lessons and exercises given in this book have long been used by us in manuscript form.

They are now published for the convenience of our own pupils and the benefit of others.

The knowledge contained in "Aids to Correct and Effective Elocution" has been drawn from many sources and tested by actual use for years among our numerous pupils.

Its teachings will be found to be in harmony with those of the best authorities on the art of expression.

The Readings and Recitations have all been chosen on account of their suitability for elocutionary practice. Those that are not of high literary merit have recommended themselves to us by the scope they give for vocal and physical expression.

CONTENTS.

AIDS TO CORRECT AND EFFECTIVE ELOCUTION.

SELECTED READINGS AND RECITATIONS FOR PRACTICE.

CONTENTS.

PAGE

The Convict Ship,	Mrs. Ann T. Stephens,	154
The Lady of Castlenore,	T. B. Aldrich,	161
The King and the Child,	Eugene J. Hall,	164
The Children's Crusade (With Music),	Longfellow,	166
Heliotrope,	From "Acta Columbiana,"	173
The Last Ride,	Mary A. P. Stansbury,	175
The Galley Slave,	Henry Abbey,	181
The Sea Breeze and the Scarf,	Ella Wheeler Wilcox,	183
The Mantle of St. John de Matha,	Whittier,	184
Sleep,	Mrs. Browning,	187
The Legend of St. Mark,	Whittier,	189
Scenes from King Richard III.,	Shakespeare,	191
Mother and Poet,	Mrs. Browning,	201
Scene in a Tenement House,		205
The Female Martyr,	Whittier,	209
Pancratius,	Eleanor C. Donnelly,	212
Catherine and Griffith,	Shakespeare,	216
Gualberto's Victory,	Eleanor C. Donnelly,	218
Queen Archidamia,		222
Prologue to Cato,	Pope,	225
Cato's Senate,	Addison,	226
The Battle of Waterloo,	Byron,	233
The Bridal of Malahide.	Gerald Griffin,	235
The Mourners,	E. Cooke,	238
An Order for a Picture,	Alice Cary,	240
The Rosary of My Tears,	Father Ryan,	244
Labor,	Thomas Carlyle,	245
The Rustic Bridal,	Longfellow,	246
The Bravest Battle that ever was Fought,	Joaquin Miller,	252
The Wives of Weinsberg,	Gottfried August Bürger,	253
Little Joe,	Jennie Woodville	256
The Merchant and the Book-Agent,		268
Brer Rabbit and the Tar Baby,	Joel Chandler Harris,	271
A Battle of Words,	Richard R. Madden,	279
Sam Weller's Valentine,	Dickens,	283
The Indians and the Mustard,		289
Caudle's Wedding Day,	Douglas Jerrold,	291

viii

CONTENTS.

AIDS

TO

CORRECT AND EFFECTIVE ELOCUTION.

GESTURE.

GESTURE is a universal language.

It is usually defined as the various postures and motions of the body.

We would define gesture as *the body's attempt to give expression to the thought.*

All laws for gesture must rest upon the primary one of *Correspondence.*

Gestures, or movements of the body, should be made with *precision, ease,* and *harmony*—in a word, with *grace.*

Although grace is defined as the union of ease, precision, and harmony, the student is cautioned against bringing any one of these into undue prominence.

We have seen ease degenerate into vulgarity, precision into pedantry, and harmony into affectation.

Gesture must always precede speech. The sense is not in the words ; it is in the inflection and gesture.

Let as much expression as possible be given to

7

the face. A gesture made by the hand is wrong when not justified in advance by the face.

Gestures should not be multiplied; we are moved by only one sentiment at a time : hence it is useless to multiply gestures.

The lines of gesture whether referring to objects or ideas are identical.

As gesture is a muscular movement, all the *directions* to gain flexibility should be carefully followed and the *exercises* faithfully practised.

Delaumosne gives six laws of gesture, viz.: *Priority, Retroaction, The Opposition of Agents, Unity, Stability,* and *Rhythm.*

PRIORITY.—This law should be carefully observed. The expression of the face should precede gesture, and gesture should precede speech.

RETROACTION.—This law is founded upon the fact that "every object of agreeable or disagreeable aspect which surprises us makes the body *recoil.* The degree of reaction should be proportionate to the degree of emotion caused by the sight of the object."

THE OPPOSITION OF AGENTS.—This law teaches that "simultaneous movement must be made in opposition." Successive movement should be parallel. It is the law of equilibrium. *In ancient art this law is always observed.*

UNITY.—This law has relation to the number of gestures. Delaumosne says : "There must be unity in everything ; but a rôle may be condensed in two or three traits ; therefore a great number of gestures is not necessary. Let it be carefully noted: the expression of the face should make the gesture of the arms forgotten."

STABILITY.—This law refers to the duration of gesture. The suspension or prolongation of movement is one of the great sources of effect.

RHYTHM.—This law teaches that gesture is rhythmic through its movement, more or less slow, or. more or less rapid.

"The rhythm of gesture is proportional to the mass to be moved." This law is based upon the vibration of the pendulum. Great levers have slow movements, small agents more rapid ones.

"In proportion to the depth and majesty of the emotion is the deliberation and slowness of the motion ; and, *vice versa*, in proportion to the superficiality and explosiveness of the *emotion* will be the velocity of its expression in *motion*."

To facilitate expression we give a system of notation similar to that invented and used by Mr. Austin in his "Chironomia," published in 1806, and than which no better work on the *technique of gesture* has since appeared :

d. f., descending front.	*p.,* prone.
d. o., descending oblique.	*v.,* vertical.
d. l., descending lateral.	*i.* or *ind.,* index finger.
d. o. b., descending oblique backwards.	*upl.,* uplifted.
	par., parallel.
h. f., horizontal front.	*cli.,* clinched.
h. o., horizontal oblique.	*cla.,* clasped.
h. l., horizontal lateral.	*ap.,* applied.

h. o. b., horizontal oblique
  backwards.
r. h., right hand.
l. h., left hand.
b. h., both hands.
s., supine.

fol., folded.
cro., crossed.
prep., preparation.
rep., repeat.
sus., sustained.
tr., tremor.

THE HEAD.

The head should be well balanced, and held midway between the shoulders.

EXERCISES.

The following exercises should be practised at first very slowly. The aim of all such exercises is to enable the physical nature, the more readily and correctly, to express the mental and emotional.

1st. Holding the head easily erect, turn it, very slowly, to the right, then to the left.

2d. Bend the head forward until the chin rests upon the chest.

3d. Bend the head backward as far as possible. Care must be taken not to stretch the neck muscles too powerfully.

4th. Turn the head right oblique; hold it there while turning the face downward, then upward. Bend the head backward; hold it there while turning the face right, then left.

The head is bowed in shame.
It droops in sorrow.
It is crouched in fear.

It is set backward in pride.

It bends forward in assent.

It bends slightly in reflection.

It shows negation by a horizontal movement from side to side.

It expresses melancholy by inclining downward to the left side.

THE FACE.

The face leads in expression.

The face should reflect or mirror the sentiment to be expressed. It expands in joy, contracts in pain, and is elongated in sorrow. Of the face the eye is generally considered the most expressive feature. Interest in an object is expressed by fixing the eyes upon it; indifference, by averting them, or turning them, after a glance, away. In reflection, meditation, thoughtful consideration, and self-examination, the eyelids close over the ball in a greater or less degree, in accordance with the depth of the thoughtful consideration or the strictness of the self-examination. The eyes express depreciation of an object by first glancing askance at it, and then partly closing.

Although the eye is usually considered the most expressive feature of the face, close observation will convince that it really needs the expressions or gestures of the mouth and nose, and sometimes also of the upper part of the face, the forehead and eyebrows, to translate its language.

The eye may be said to lead the face in expression as it indicates. To express disdain and

disgust it needs the assistance of the nose and
mouth.

To express astonishment the eyes are wide open
and the brows raised.   Perplexity in thought is
expressed by a slight frown.   All aggressive gestures are accompanied by the knitting of the
brows.   The eyes are raised to express faith,
hope, and love.   They are cast down in modesty.

THE HAND.

The hand with the aid of the face can express
every mood, translate every language.

The hand is an assistant to the expression the
face has already given.   In a proper carriage of
the hand the fingers group themselves thus : the
two middle fingers are held together slightly bent
inwards.   The index-finger and the little finger
are separated slightly from the two middle fingers,
the index nearly straight, the little finger slightly
curved.   The thumb is held nearly on a line with
the index-finger.

The clinched hand signifies conflict.

The hand closed with the thumb at the side of
the first finger signifies power.

The hand partly opened from the clinched fist,
with the fingers somewhat apart and the first
joint bent, signifies a convulsive state.   Execration
is expressed in the same manner, only the hand
is more opened.

The hand opened to full extent expresses ex-
altation.   The hand opened to full extent with
the fingers wide apart signifies exasperation.

We use the tips of the fingers when illustrating fine distinctions or urging nice critical points. The hand prone expresses superposition—one thing placed on or lying upon another.

While the supine hand expresses the naked truth, the prone clothes the thought with some repressive emotion, as grief, sadness, or any sentiment of a grave, solemn, and subdued character.

The supine hand permits, the prone rejects; the supine impels, the prone restrains.

The supine is open, frank, genial; the prone is aversive. The supine hand indicates nearness; the prone, distance.

Directive gestures are necessary when the words *Lo! yonder, this, that, behold!* etc., are used.

In gesticulation it is of vital importance that *the whole arm initiate the movement.* Perfect relaxation of the muscles of the arm should precede this initial movement.

EXERCISES.

1st. Relax the muscles of the arms, then turn the trunk of the body, letting the arms flop right and left. Extend the right arm horizontal front. Move the hand (vertically) up and down; move it horizontally right and left.

2d. Rotate the arm. Rotate the hand, the centre of motion being the elbow.

The above exercises should be practised until the student has acquired flexibility of the muscles.

Without flexibility there can be no grace of gesture.

THE ELBOW.

"The elbow turned outward signifies strength, power, audacity, domination, arrogance, abruptness, activity. The elbow drawn inward signifies impotence, fear, subordination, humility, passiveness, poverty of spirit."

CARRIAGE OF THE BODY.

An easy and correct bearing is of the first importance.

Care should be taken to hold the abdomen in, the chest up, and the shoulders backward and downward.

The carriage or bearing of the speaker should correspond to and be in harmony with the sentiment he is uttering.

EXERCISES.

The following exercises should be practised until the pupil is able to "stand still with ease and move with grace:"

1st. Stand easily erect, with the weight of the body resting equally upon the feet. Carefully hold the abdomen *up* by the muscles of the thorax and the back, and *in* by the abdominal muscles.

2d. With the weight of the body resting upon either hip, carry the leg that is then free forward with the knee slightly bent.

3d. Standing in the first position, place one foot behind, resting the weight upon it, then bend the backward knee until it rests upon the floor. Incline the trunk of the body forward and the head backward. To rise, shift the weight of the body to the advanced leg.

WALKING.

In walking, the ball of the foot should meet the ground first.

Ladies can walk with or without heels provided they have learned to walk.

No shoes which interfere with the hinge-like movement necessary in good walking should be worn.

"The elastic step of youth," of which poets love to sing, is really the step of every lady, old or young, who has acquired the difficult art of walking.

To walk well one must first *breathe well.*

The chest should be held high and the abdomen drawn backward. If the chest is held high, the shoulders will naturally fall backward and downward. If too much effort be made to draw the shoulders backward and downward, one is apt to throw the abdomen outward.

In walking, the legs should swing from the hip-joints—or, to speak more elegantly, "the impulse should be given from the thigh," and the toes be turned slightly outward.

EXERCISES.

1st. Standing in the first position (i.e., with the heels together and the weight of the body resting equally upon the feet), lift the right leg and swing it from the hip-joint.

2d.   Standing upon the right foot, practise the same with the left leg.

3d.   Standing in the first position, rise slowly upon the toes, then resume first position.

4th.   Advance right foot and practise hinge-like movement with left.

5th.   Advance left foot and practise hinge-like movement with right.

BOWING.

The bow, which has taken the place of the deep courtesy, is made by an inclination of the body from the ankles, and signifies "I am at your service."

The courtesy—which is still used in our academies—is made by putting one foot behind, resting the weight upon it, and then bending the backward knee (the forward will bend also), at the same time relaxing the muscles of the upper part of the body, and bowing the head. In drawing up the form and returning to position, the movement should be very slow, as this kind of bow signifies reverence. Bowing, kneeling, and seating one's self with grace are accomplished by observing the law of *poise*, or "Opposition of Agents." The law consists in placing the acting levers in opposition, and thus realizing *equilibrium*. And as Delaumosne says most truly, "All that is in equilibrium is harmonized."

We should make our bow to the audience before, not while, uttering the title of our Reading or Recitation.

TO SEAT ONE'S SELF.

Standing in the first position before a seat, place one foot backward and courtesy into the seat. Carefully observe that the *feet bear the weight of the body* until it reaches the seat.

To rise, *press* upon the feet. The natural impulse is to assist one's self with the hands; this must be avoided.

BREATHING.

As voice is formed by the breath setting in vibration the lips of the glottis, it follows that the management of the breath is of the first importance.

As all the air entering the lungs and all the breath leaving them must pass through the larynx, *inspiration* and *expiration* must alternate.

In deep breathing the cavity within the chest is increased by the descent of the diaphragm, also by the expansion of the ribs.

In inspiration the abdomen slightly protrudes. In expiration the abdomen falls inward.

When the lungs are entirely filled they will be found to expand the *back* as well as the chest.

All *impediments to free respiration* are to be avoided. The *throat*, *chest*, and *abdomen* must be left free action.

When the lungs are kept *well supplied* with air vocal efforts are healthy and unfatiguing.

In reading and reciting, the pupil is cautioned against allowing the lungs to become exhausted. We most earnestly recommend a "silent pause" to replenish them—or rather, to allow the atmospheric pressure to do so.

The tendency to drop the voice at the end of each line in poetry is often the result of neglecting this "silent pause." In effective and expressive reading and recitation silence is indeed golden.

Respiration should be *inaudible.*

We have two channels for respiration, the nostrils and the mouth.

Audible respiration is caused by a contraction of these air passages.

It is a disputed point as to which of the channels for respiration should be preferred. We recommend *both*, if the organs be in a *perfectly sound and healthy condition*, and the atmosphere *pure*. Should the student be delicate, or the atmosphere impure, we advise inhaling through the nostrils—which have been called "nature's filter."

EXERCISES.

1st. Deep Breathing.—Place the hands upon the hips. Exhaust the lungs, then fill them very slowly and thoroughly, retain the breath a short time, and then as slowly emit it.

2d. Inflate the lungs, then empty them suddenly, giving the sound of the aspirate *h*.

THE VOICE.

The human voice is the most wonderful of all musical instruments, and the most satisfactorily expressive when "the soul attunes the instrument to the theme." The voice has three properties —*Pitch, Force,* and *Quality;* and it is to the cultivation of these three properties of the voice that the best efforts of the student should be directed.

Indeed, in a technical training of the voice, all the artifices of the Elocutionist are founded upon the three properties of the voice, *Pitch, Force,* and *Quality.* What the elder Garcia said of the singing voice—"Its beauty constitutes ninety-nine hundredths of the power of the singer"—is equally true of the speaking voice. For this reason we disapprove of giving to young students recitations requiring certain faulty qualities of the voice. The aim of the teacher of vocal culture should be *to preserve and improve the natural beauty of the voice.* L'Abbé Delaumosne says: "It is through the *voice* we please an audience. If we have the ear of an auditor, we easily win his mind and heart. The voice is a mysterious hand which touches, envelops, and caresses the heart."

PITCH.

Pitch relates to the elevation or depression of the voice.

For direction five degrees of Pitch are distinguished: *Very Low, Low, Middle, High,* and *Very High.* Although but five degrees of pitch are given for direction in Elocution, the student will note that "the degrees range through the entire compass of the voice."

Middle Pitch is used in unimpassioned styles.

High Pitch is employed to express exultation, joy, etc.

Very High Pitch is used to express the extremes of joy and grief, and is employed in calling.

Low Pitch expresses solemnity, awe, reverence, and sorrow.

Very Low Pitch expresses extreme awe, solemnity, reverence, dread, etc.

A harmony of pitch must be maintained between closely related parts of a sentence, such as subject and verb, verb and object, especially when they are separated by intervening clauses.

FORCE.

Force relates to the degree of energy with which a sound is given.

For direction in Elocution five degrees of Force are given: *Weak, Subdued, Moderate, Energetic,* and *Vehement.* The following recitation, " King Henry the Second at the Tomb of King Arthur," is an excellent exercise on Force. The first part is given with Moderate Force; Energetic Force is used in the fourth verse; Vehement in the fifth; while the concluding verses are given with Subdued and Weak, or Gentle, Force.

KING HENRY THE SECOND AT THE TOMB OF KING ARTHUR.

THE tourney past, in festival
  Baron and knight were met :
Last pomp it was that graced the hall
  Of great Plantagenet ;
A Prince for valor praised by all,
  More famed for wisdom yet.

The board rang loud with kingly cheer :
  Light jest, and laugh, and song
Rang swiftly round from peer to peer ;
  Alone on that gay throng
The harper looked with eye severe,
  The while in unknown tongue

A mournful dirge abroad he poured ;
  Sad strains, forlorn, and slow :
Poor wreck of music prized and stored
  Long centuries ago
On British hills ere Saxon sword
  Had stained as yet their snow.

" Strike other chords !" the monarch cried ;
  " Whate'er thy words may be,
They sound the dirge of festal pride :
  Warriors, not monks, are we !
The melodies to grief allied
  No music make for me."

Louder and louder pealed the strain,
  More wild, and soul-entrancing :
Picturing now helmets cloven in twain,
  Now swords like meteors glancing ;
Now trampling hosts o'er hill and plain
  Retreating and advancing.

The bard meanwhile with cold, stern air,
    Looked proudly on the proud,
Fixing unmoved a victor's stare
    On that astonished crowd—
Till all the princes gathered there
    Leaped up, and cried aloud: "

" What man, what chief, what crownèd head
    Eternal heir of fame,
Of all that live, or all the dead,
    This praise shall dare to claim ?"
Then rose that British bard, and said,
    " King Arthur is his name."

"What sceptre grasped King Arthur's hand?"
    " The sceptre of this Isle."
" What nations bled beneath his brand ?"
    " The Saxon foe erewhile."
" His tomb ?" was Henry's next demand—
    " He sleeps in yonder pile."

Forth went the King with all his train,
    At the mid hour of night ;
They paced in pairs the silent plain
    Under the red torch-light.
The moon was sinking in her wane,
    The tower yet glimmered bright.

The grave they found ; their swift strokes fell,
    Piercing dull earth and stone.
They reached ere long an oaken cell,
    And cross of oak, whereon
Was graved, " Here sleeps King Arthur well,
    In the Isle of Avalon."

The mail on every knightly breast,
  The steel at each man's side,
Sent forth a sudden gleam : each crest
  Bowed low its plumèd pride :
Down o'er the coffin stooped a priest—
  But first the monarch cried,

" Great king ! in youth I made a vow
  Earth's mightiest son to greet ;
His hand to worship ; on his brow
  To gaze ; his grace entreat.
Therefore, though dead, till noontide thou
  Shalt fill my royal seat !"

Away the massive lid they rolled—
  Alas ! What found they there ?
No kingly brow, no shapely mold ;
  But dust where such things were.
Ashes o'er ashes, fold on fold,
  And one bright wreath of hair.

Then Henry lifted from his head
  The Conqueror's iron crown :
That crown upon that dust he laid,
  And knelt in reverence down,
And raised both hands to heaven, and said,
  " Thou, God, art King alone !"

" Lie there, my crown, since God decrees
  This head a couch as low !
What am I better now than these
  Six hundred years ago ?
Henceforth all mortal pageantries
  I count an idle show."—*Aubrey de Vere.*

QUALITY.

Quality relates to the kind of tone.

Broadly speaking, there are but two kinds of tone—*the pure* and *the impure.*

The pure tone is that in which all the breath is vocalized. The impure, or aspirate, is the tone in which only a part of the breath is vocalized.

The pure tone enlarged and intensified becomes the Orotund—"a full round tone used to express grand thoughts or to picture sublime objects."

The Guttural, so called because the resonance is chiefly in the guttur, or throat, is employed to express intense hatred, scorn, and rage.

While the voice is being developed, we do not advise prolonged practice of gutturals.

PITCH, FORCE, AND QUALITY.

Scientific investigation reports that *Force* depends on the amplitude of the air waves set in motion by some mass of matter in vibration; *Pitch* upon the number of vibrations given forth in a second of time. The fewer the vibrations the lower the pitch of the sound; the greater the number of vibrations in each second the higher the pitch. *Quality*, or Timbre, depends upon the forms of the air waves sent out from the vibrating body.

TIME.

Time is the rate of utterance.

It is not a property of the voice, but simply relates to its continuance for a longer or shorter period.

For practice we give *Moderate,* or *Common, Time, Quick Time, Very Quick Time, Slow Time,* and *Very Slow Time.*

The rate of utterance must, of course, depend upon the sentiment or emotion to be expressed.

Unimpassioned discourse and simple description take Moderate Time; animated descriptions, Quick Time; violent passion, Very Quick Time; tenderness takes Slow Time; solemnity, Very Slow Time.

STRESS.

The *manner* in which Force is applied forms an important part of " vocal technique."

Six kinds of Stress are usually given:

>   The Radical Stress.
>   The Terminal Stress.
>   The Median Stress.
>   The Compound Stress.
>   The Tremor.
>   The Thorough Stress.

In the Radical, as the name implies, the greatest stress is given at the *beginning* of the sound, and in the Terminal at the *end.*

The Median Stress is given by gradually increasing and as gradually diminishing the sound.

The Compound Stress is simply the Radical and Terminal united.

The Tremor—sometimes called the Intermittent Stress—is a trembling of the voice.

In the Thorough Stress the Force is sustained.

ON THE USE OF THE SIX KINDS OF STRESS.

The Radical Stress is used to express positive and decisive convictions. "England's Treatment of Ireland" affords an opportunity for the application of this kind of Stress.

ENGLAND'S TREATMENT OF IRELAND.

WHAT is the case of Ireland at this moment? Have the gentlemen considered that they are coming into contact with a nation? This, if I understand it, is one of the golden moments of our history, one of those opportunities which may come, may go, but which rarely return, or, if they return, return at long intervals, and under circumstances which no man can forecast. There have been such golden moments even in the tragic history of Ireland. There was such a golden moment in 1795 during the mission of Lord Fitzwilliam, and at that moment it is historically clear that the Parliament of Grattan was on the point of solving the Irish problem. The two great knots of that problem were Catholic emancipation and reform of Parliament. The cup was at her lips and she was ready to drink it, when the hand of England rudely and ruthlessly dashed it to the ground in obedience to the wild and dangerous intimation of an Irish faction.

There has been no great day of hope for Ireland, no day when you might hope completely and definitely to end the controversy, till now, after more than ninety years. The long periodic time

has at last run out, and the star has again
mounted up into the heavens. What Ireland
was doing for herself in 1798, we at length have
done. The Roman Catholics have been emancipated—emancipated after a woful disregard of
solemn promises through twenty-nine years,
emancipated slowly, sullenly, not from good-
will, but from abject terror, with all the fruits
and consequences that will follow that method
of legislation. The second problem has been
also solved: the representation of Ireland re-
formed; the franchise given to her with the read-
justment with a free and open hand. That gift
of franchise was the last act required to make the
success of Ireland in her final effort absolutely
sure. We have given Ireland a voice and we
must listen to what she says. We must all listen,
both sides, both parties. Ireland stands at your
bar expectant, hopeful, almost suppliant. Her
words are the words of truth and soberness.
She asks blessed oblivion of the past, and in that
oblivion our interest is a deeper interest than
hers. Go into the length and breadth of the
world, search the literature of all countries, and
find if you can a single voice, a single book in
which the conduct of England towards Ireland
is anywhere treated except with profound and
bitter condemnation.

Are these the traditions by which we are ex-
horted to stand? No! They are a sad exception to the glory of our country. They are
more than a black blot upon the pages of its his-

tory. And what we want to do is to stand by the
traditions of which we are the heirs in all matters
except our relations with Ireland. So we treat
our traditions; so we hail the demand of Ireland
for a blessed oblivion of the past. She asks also
a boon for the future, and that will be a boon to
us in respect to honor no less than to her in respect to happiness, prosperity, and peace.

William E. Gladstone.

The Terminal Stress is used to express scorn,
defiance, and revenge when used violently, and
peevishness and impatience when used lightly.
The Terminal Stress prevails in "The Semi-
nole's Defiance."

THE SEMINOLE'S DEFIANCE.

BLAZE! with your serried columns,
  I will not bend the knee!
The shackle ne'er again shall bind
  The arm which *now* is free :
I have mailed it with the thunder
  When the tempest muttered low:
And where it falls ye well may dread
  The lightning of its blow!

I've scared ye in the city,
  I've scalped ye on the plain;
Go, count your chosen where they fell
  Beneath my leaden rain.
I scorn your proffered treaty,
  The pale-face I defy;
Revenge is stamped upon my spear,
  And "*Blood !*" my battle cry.

Some strike for hope of booty,
   Some to defend their all—
I battle for the joy I have
   To see the white man fall!
I love, among the wounded,
   To hear his dying moan,
And catch, while chanting at his side,
   The music of his groan.

Ye've trailed me through the forest,
   Ye've tracked me o'er the stream,
And, struggling through the ever-glade,
   Your bristling bayonets gleam :
But I stand as should the warrior,
   With his rifle and his spear :
The scalp of vengeance still is red,
   And warns ye—*Come not here !*

Think ye to find my homestead ?
   I gave it to the fire !
My tawny household do ye seek ?
   I am a childless sire !
But should ye crave life's nourishment,
   Enough I have and good :—
I live on hate—'tis all my bread,
   Yet light is not my food.

I loathe you with my bosom,
   I scorn you with mine eye—
And I'll taunt you with my latest breath,
   And fight you till I die!

I ne'er will ask you quarter,
   And I ne'er will be your slave :
But I'll swim the sea of slaughter,
   Till I sink beneath its wave.

                        G. W. Patten.

The Median Stress is used in tranquil emotions. It is most appropriate in poetic description. The movement is smooth and gliding. "A Forest Hymn" should be given with Median Stress.

A FOREST HYMN.

THE groves were God's first temples. Ere man
    learned
To hew the shaft, and lay the architrave,
And spread the roof above them—ere he framed
The lofty vault, to gather and roll back
The sound of anthems, in the darkling wood,
Amid the cool and silence, he knelt down,
And offered to the Mightiest solemn thanks
And supplication. For his simple heart
Might not resist the sacred influence
Which, from the stilly twilight of the place,
And from the gray old trunks that high in heaven
Mingled their mossy boughs, and from the sound
Of the invisible breath that swayed at once
All their green tops, stole over him, and bowed
His spirit with the thought of boundless power
And inaccessible majesty.
                  Father, Thy hand
Hath reared these venerable columns, Thou
Didst weave this verdant roof. Thou didst look
    down

Upon the naked earth, and, forthwith, rose
All these fair ranks of trees. They, in Thy sun,
Budded, and shook their green leaves in Thy
    breeze,
And shot toward heaven.
                      Grandeur, strength, and grace
Are here to speak of Thee. This mighty oak—
By whose immovable stem I stand and seem
Almost annihilated—not a prince,
In all that proud old world beyond the deep,
E'er wore his crown as loftily as he
Wears the green coronal of leaves with which
Thy hand has graced him. Nestled at his root
Is beauty, such as blooms not in the glare
Of the broad sun. That delicate forest flower,
With scented breath and look so like a smile,
Seems, as it issues from the shapeless mould,
An emanation of the indwelling Life,
A visible token of the upholding Love,
That are the soul of this great universe.
   My heart is awed within me when I think
Of the great miracle that still goes on,
In silence, round me—the perpetual work
Of Thy creation, finished, yet renewed
Forever. Written on Thy works I read
The lesson of Thy own eternity.
Lo! all grow old and die—but see again,
How on the faltering footsteps of decay
Youth presses—ever gay and beautiful youth
In all its beautiful forms. These lofty trees
Wave not less proudly that their ancestors
Moulder beneath them. Oh, there is not lost

One of earth's charms : upon her bosom yet,
After the flight of untold centuries,
The freshness of her far beginning lies
And yet shall lie.   Life mocks the idle hate
Of his arch-enemy Death—yea, seats himself
Upon the tyrant's throne—the sepulchre,
And of the triumphs of his ghastly foe
Makes his own nourishment.   For he came forth
From thine own bosom, and shall have no end.
Oh ! let me often to these solitudes
Retire, and in Thy presence reassure
My feeble virtue.   Here its enemies,
The passions, at Thy plainer footsteps shrink
And tremble and are still.   O God ! when Thou
Dost scare the world with tempests, set on fire
The heavens with falling thunderbolts, or fill,
With all the waters of the firmament,
The swift dark whirlwind that uproots the woods
And drowns the villages ; when, at Thy call,
Uprises the great deep and throws himself
Upon the continent, and overwhelms
Its cities—who forgets not, at the sight
Of these tremendous tokens of Thy power,
His pride, and lays his strifes and follies by ?
Oh, from these sterner aspects of Thy face
Spare me and mine, nor let us need the wrath
Of the mad unchained elements to teach
Who rules them.   Be it ours to meditate,
In these calm shades, Thy milder majesty,
And to the beautiful order of Thy works
Learn to conform the order of our lives.

William Cullen Bryant.

The Compound Stress is used to render a mingling of emotions, as surprise and anger, scorn and contempt. "The Chameleon" gives the student an opportunity to practise this kind of stress.

THE CHAMELEON.

Oft has it been my lot to mark
A proud, conceited, talking spark,
With eyes that hardly served at most
To guard their master 'gainst a post;
Yet round the world the blade has been,
To see whatever could be seen.
Returning from his finished tour,
Grown ten times perter than before,
Whatever word you chance to drop
The travelled fool your mouth will stop:
" Sir, if my judgment you'll allow,
I've seen—and sure I ought to know."
So begs you'd pay a due submission,
And acquiesce in his decision.

Two travellers of such a cast,
As o'er Arabia's wilds they passed,
And on their way, in friendly chat,
Now talked of this, and then of that;
Discoursed awhile, 'mongst other matter,
Of the chameleon's form and nature.
" A stranger animal," cries one,
" Sure never lived beneath the sun:
A lizard's body lean and long,
A fish's head, a serpent's tongue,

Its foot with triple claw disjoined,
And what a length of tail behind !
How slow its pace ! and then its hue,—
Who ever saw so fine a blue ?"

↘ " Hold there !" the other quick replies ;
" 'Tis green : I saw it with these eyes,
As late with open mouth it lay
And warmed it in the sunny ray ;
Stretched at its ease the beast I viewed,
And saw it eat the air for food."
   " I've seen it, sir, as well as you,
And must again affirm it blue ;
At leisure I the beast surveyed,
Extended in the cooling shade."
      " 'Tis green, 'tis green, sir, I assure ye."
" Green !" cries the other, in a fury:
" Why, sir, d'ye think I've lost my eyes ?"
" 'Twere no great loss," the friend replies ;
" For if they always serve you thus,
You'll find them but of little use."
   So high at last the contest rose,
From words they almost came to blows,—
When luckily came by a third :
To him the question they referred,
And begged he'd tell them, if he knew,
Whether the thing were green or blue.
   " Sirs," cries the umpire, "cease your pother ;
The creature's neither one nor t'other.
I caught the animal last night,
And viewed it o'er by candle-light :

I marked it well; 'twas black as jet—
You stare; but, sirs, I've got it yet,
And can produce it."—" Pray, sir, do ;
I'll lay my life the thing is blue."
" And I'll be sworn that when you've seen
The reptile you'll pronounce him green."

" Well, then, at once to ease the doubt,"
Replies the man, " I'll turn him out ;
And when before your eyes I've set him,
If you don't find him black I'll eat him."

He said,—and full before their sight
Produced the beast, and lo ! 'twas white.
Both stared ; the man looked wondrous wise.
" My children," the chameleon cries,—
Then first the creature found a tongue,—
" You all are right and all are wrong :
When next you talk of what you view,
Think others see as well as you,
Nor wonder if you find that none
Prefers your eyesight to his own."

James Merrick.

The Tremor is used to express excessive emotion. The voice will tremble or shake in any great emotion of whatever nature. In reciting the part of Arthur in the " Scene from King John" this stress is used.

SCENE FROM KING JOHN.
Act IV., Scene I.

HUBERT. ARTHUR. ATTENDANTS.

Hub. Heat me these irons hot; and look you
    stand

Within the arras:[1] when I strike my foot
Upon the bosom of the ground, rush forth,
And bind the boy which you shall find with me
Fast to the chair: be heedful: hence, and watch.

   1 *Attend.* I hope your warrant will bear out the
     deed.

   Hub. Uncleanly scruples! fear not you: look
     to't.—   [*Exeunt Attendants.*

Young lad, come forth; I have to say with you.

Enter ARTHUR.

   Arth. Good morrow, Hubert.
   Hub.   Good morrow, little Prince.
   Arth. As little prince, having so great a title
To be more[2] prince, as may be. You are sad.
   Hub. Indeed, I have been merrier.
   Arth.   Mercy on me!
Methinks nobody should be sad but I:
Yet, I remember, when I was in France,
Young gentlemen would be as sad as night,
Only for wantonness. By my christendom,[3]
So I were out of prison, and kept sheep,
I should be merry as the day is long;
And so I would be here, but that I doubt[4]

---

   [1] *Arras* were the hangings or tapestries with which rooms
were lined, before the days of plastering. To keep them
from being rotted by the damp, they were hung on frames,
far enough from the walls to admit of a person's hiding behind them.

   [2] *More* for *greater.*

   [3] *Christendom* for *christening* or *baptism.* The usage was
common.

   [4] *Doubt* in the sense of *fear* or *suspect;* a frequent usage.—
Practises, in the next line, is *contrives, plots,* or *uses arts.*

My uncle practises more harm to me:
He is afraid of me, and I of him:
Is it my fault that I was Geffrey's son?
No, indeed, is't not; and I would to Heaven
I were your son, so you would love me, Hubert.

    Hub. [*Aside.*] If I talk to him, with his inno-
        cent prate
He will awake my mercy, which lies dead:
Therefore I will be sudden and dispatch.

    Arth. Are you sick, Hubert? you look pale to-
        day:
In sooth,[1] I would you were a little sick,
That I might sit all night and watch with you:
I warrant I love you more than you do me.

    Hub. [*Aside.*] His words do take possession of
        my bosom.—
Read here, young Arthur.—   [*Showing a paper.*
        [*Aside.*] How, now, foolish rheum![2]
Turning dispiteous[3] torture out of door!
I must be brief, lest resolution drop
Out at mine eyes in tender womanish tears.—
Can you not read it? is't not fairly writ?

    Arth. Too fairly, Hubert, for so foul effect:
Must you with hot irons burn out both mine
        eyes?

    Hub. Young boy, I must.

    Arth.   And will you?

    Hub.   And I will.

---

[1] In *truth* or *truly*.

[2] *Rheum* for *tears.*

[3] *Dispiteous* for *unpiteous*, that is, *pitiless.*—In the next line,
brief is *quick, prompt,* or *sudden.*

Arth. Have you the heart? When your head
    did but aehe,
I knit my handkercher about your brows,—
The best I had, a princess wrought it me,—
And I did never ask it you again;
And with my hand at midnight held your head;
And, like the watchful minutes to the hour,[1]
Still and anon cheer'd up the heavy time,
Saying, *What lack you?* and, *Where lies your grief?*
Or, *What good love may I perform for you?*
Many a poor man's son would have lain still,
And ne'er have spoke a loving word to you;
But you at your sick service[2] had a prince.
Nay, you may think my love was crafty love,
And call it cunning; do, an if[3] you will:
If Heaven be pleased that you must use me ill,
Why, then you must. Will you put out mine
      eyes?
These eyes that never did nor never shall
So much as frown on you?
   Hub.   I've sworn to do it;
And with hot irons must I burn them out.
   Arth. Ah, none but in this iron age would do it!
The iron of itself, though heat[4] red-hot,

[1] That is, as the minutes watch over, or mark, the progress
or passage of the hour. A pretty way of expressing a minute
and sedulous attention.—"*Still* and anon," in the next line, is
the same as our "*ever* and anon."

[2] *Sick service* is of course merely an instance of what is called
transferred epithet: service *done to the sick.*

[3] *An if* is an old reduplication much used in the Poet's
time. So we have an, or *if*, or *an if*, used indifferently.

[4] *Heat* for *heated.*

Approaching near these eyes, would drink my
      tears,
And quench its fiery indignation
Even in the water of mine innocence ;
Nay, after that, consume away in rust,
But for containing fire to harm mine eyes.
Are you more stubborn-hard than hammer'd
      iron ?
An if an angel should have come to me,
And told me Hubert should put out mine eyes,
I would not have believed him,—no tongue but
      Hubert's.
    Hub. Come forth!   [*Stamps.*

    Re-enter ATTENDANTS, *with cord, irons, etc.*

Do as I bid you do.
    Arth. O, save me, Hubert, save me! my eyes
    are out
Even with the fierce looks of these bloody men.
    Hub. Give me the iron, I say, and bind him
    here.
    Arth. Alas, what need you be so boisterous-
    rough ?
I will not struggle, I will stand stone-still.
For Heaven-sake, Hubert, let me not be bound !
Nay, hear me, Hubert ! drive these men away,
And I will sit as quiet as a lamb ;
I will not stir, nor wince, nor speak a word,
Nor look upon the iron angerly :
Thrust but these men away, and I'll forgive you,
Whatever torment you do put me to.
    Hub. Go, stand within ; let me alone with him.

1 *Attend.* I am best pleased to be from such a
    deed.   [*Exeunt Attendants.*
Arth. Alas, I then have chid away my friend!
He hath a stern look, but a gentle heart:
Let him come back, that his compassion may
Give life to yours.
    Hub.   Come, boy, prepare yourself.
    Arth. Is there no remedy?
    Hub.   None, but to lose your eyes.
    Arth. O Heaven, that there were but a mote in
        yours,
A grain, a dust, a gnat, a wandering hair,
Any annoyance in that precious sense!
Then, feeling what small things are boisterous [1]
    there,
Your vile intent must needs seem horrible.
    Hub. Is this your promise? go to, hold your
        tongue.
    Arth. Hubert, the utterance of a brace of
        tongues
Must needs want pleading for a pair of eyes:
Let me not hold my tongue, let me not, Hubert;
Or, Hubert, if you will, cut out my tongue,
So I may keep mine eyes: O, spare mine eyes,
Though to no use but still to look on you!
Lo, by my troth, the instrument is cold,
And would not harm me.
    Hub.   I can heat it, boy.
    Arth. No, in good sooth; the fire is dead with
        grief,

---

[1] *Boisterous* was used much more variously than at present;
as a common antithesis to *gentle*, and so for *rough, rude, violent*, etc.

Being create for comfort, to be used
In undeserved extremes:[1] see else yourself;
There is no malice burning in this coal;
The breath of heaven hath blown his spirit out,
And strew'd repentant ashes on his head.

   Hub. But with my breath I can revive it, boy.

   Arth. An if you do, you will but make it blush,
And glow with shame of your proceedings, Hu-
      bert:
Nay, it perchance will sparkle in your eyes;
And, like a dog that is compell'd to fight,
Snatch at his master that doth tarre[2] him on.
All things that you should use to do me wrong
Deny their office: only you do lack
That mercy which fierce fire and iron extend,
Creatures of note for mercy-lacking uses.

   Hub. Well, see to live; I will not touch thine
      eyes
For all the treasure that thine uncle owes:
Yet am I sworn, and I did purpose, boy,
With this same very iron to burn them out.

   Arth. O, now you look like Hubert! all this
      while
You were disguisèd.

   Hub.   Peace; no more. Adieu.
Your uncle must not know but you are dead;
I'll fill these doggèd spies with false reports:

---

[1] *Extremities*, or *extreme severities*, that are unmerited.
Johnson paraphrases the passage as follows: "The fire, being
created not to hurt, but to comfort, is dead with grief for
finding itself used in acts of cruelty, which, being innocent,
I have not deserved."

[2] To *tarre* is to incite, to instigate, as in setting on dogs.

And, pretty child, sleep doubtless [1] and secure
That Hubert, for the wealth of all the world,
Will not offend thee.
  Arth.   O Heaven! I thank you, Hubert.
  Hub. Silence; no more: go closely [2] in with
   me:
Much danger do I undergo for thee.
         Shakespeare.

 The Thorough Stress is used for calling or proclaiming, and is necessary whenever a sustained volume of sound is required. The seventh verse of the "Battle of Hohenlinden" will afford practice of Thorough Stress.

BATTLE OF HOHENLINDEN.

On Linden, when the sun was low,
All bloodless lay the untrodden snow,
And dark as winter was the flow
 Of Iser rolling rapidly.

But Linden saw another sight,
When the drum beat at dead of night,
Commanding fires of death to light
 The darkness of her scenery.

By torch, and trumpet fast arrayed,
Each horseman drew his battle-blade;
And furious every charger neighed
 To join the dreadful revelry.

[1] *Doubtless* for *fearless*, as *doubt* for *fear* a little before.
[2] *Closely* is *secretly;* a frequent usage.

Then shook the hills with thunder riven ;
Then rushed the steed to battle driven ;
And louder than the bolts of heaven,
  Far flashed the red artillery.

And redder yet those fires shall glow
On Linden's hills of blood-stained snow ;
And darker yet shall be the flow
  Of Iser rolling rapidly.

'Tis morn,—but scarce yon lurid sun
Can pierce the war-clouds, rolling dun,
Where furious Frank and fiery Hun
  Shout in their sulph'rous canopy.

The combat deepens. On, ye brave,
Who rush to glory, or the grave !
Wave, Munich, all thy banners wave !
  And charge with all thy chivalry !

Few, few shall part where many meet !
The snow shall be their winding-sheet,
And every turf beneath their feet
  Shall be a soldier's sepulchre.
               Campbell.

TRANSITIONS.

Transitions or variations in *Pitch, Force, Quality,* and *Time* are the reader's most powerful means to interest, charm, and convince.

Due prominence or subordination is thus given to every thought, according to its relative importance.

We recommend to students whose reading is monotonous, the practice of "Transitions."

PAUSING.

There can be no correct reading, and certainly no effective reciting, without many, and sometimes long, pauses.

The law which regulates pausing is the sense. "A good reader must often pause where no grammarian could put a point."

Sterne thus satirizes the critic who would bind emotional expression by grammatical rule:

"'How did Garrick speak the soliloquy last night?' 'Oh, against all rule, my lord; most ungrammatically! Betwixt the substantive and the adjective, which should agree together in number, case, and gender, he made a breach, || stopping as if the point wanted settling. And after the nominative, which your lordship knows should govern the verb, he suspended his voice in the Epilogue a dozen times,—three seconds and three fifths by a stop-watch, my lord, each time!' 'Admirable grammarian! But in suspending his voice, was the sense suspended likewise? Did no expression of attitude or countenance fill up the chasm? Was the eye silent? Did you narrowly look?' 'I looked only at the stop-watch, my lord!' 'Excellent observer!'"

TONES AND ARTICULATIONS.

"Artistic excellence in any form of art is rarely attained in after-life if the foundation is not laid in childhood."

This is especially true of reading.

From their tenderest years pupils should be trained to *articulate* distinctly.

All intelligent reading must be preceded by comprehension.

If the reader cannot take the thought into his own mind, how can he give it to the mind of another?

Many ill effects arise from giving children language to read which they do not comprehend. If a child compose a sentence, it will never fail to give its meaning to another, and its inflections and pauses will be artistically correct. Let it understand, as thoroughly, every exercise in reading and recitation it is called upon to give, and the same artistic excellence will mark its delivery.

VOWELS AND CONSONANTS.

To develop the voice and gain control of the organs of speech, we recommend the practice of the following exercises. The first contains the extremes of vowel sounds.

EXERCISES.

1st. Open the throat, as in gaping, and give *ä* as in *father*.

Extend the lips sideways, as when smiling, and give the vowel *ē*.

Purse the lips forward, and give *ōo* as in *ooze*.

2d. Inflate the lungs thoroughly, then give each of the following sounds with the Median Stress; i.e., begin the sound softly, and gradually increase the volume of the tone to the full power of the voice, then as gradually diminish it: $\bar{e}$, $\bar{a}$, *ah*, *aw*, $\bar{o}$, $\bar{o}o$.

3d. Give the above sounds with the Radical Stress.

The student of expression is reminded that to understand and feel what is to be expressed is not sufficient. A perfect control of all the agents of expression is also required, and this control is only acquired by faithful practice.

PRONUNCIATION.

To pronounce elegantly, one must give correctly the sound of every vowel and the power of every consonant, together with the proper accentuation.

We do not think it necessary to discuss the subject of pronunciation to any great extent in this book, but only to give directions by which the pupil may acquire a thorough knowledge of this important part of elocution.

The student must appeal constantly to the best dictionaries. As, however, pronunciation is more a matter of *habit* than knowledge, we recommend students to make lists of words which they mispronounce or fail to give with sufficient distinctness of articulation.

Students should carefully observe and give the *intermediate* sound of $\dot{a}$ in such words as *last*, *past*, etc. We often hear the sound of $\breve{a}$ as in *at* inelegantly substituted. Another common fault is giving the sound of $\bar{u}$ in such words as *duke*, *news*, etc., like $\bar{o}o$.

Genius has been defined as the art of taking great pains. And certainly with industry, students of ordinary ability may succeed in acquiring an elegant pronunciation.

An excellent exercise is to select from a reading the most difficult words, and to articulate very slowly the *elementary sounds* which compose such words.

Another exercise which we have found most beneficial is to read in a whisper so as to be understood at a distance. The effort the pupil will make to compensate by distinctness of enunciation for the lack of vocal power will strengthen the organs of speech by bringing them more powerfully into play. Alexander Melville Bell, whose works are of great scientific value, declares excellence of pronunciation to depend primarily on a clear syllabication of words.

In conclusion, we beg the pupil to endeavor to realize the description given by the Rev. Mr. Austin in his "Chironomia" of the words of a good speaker, which he declares to be like "beautiful coins newly issued from the mint, deeply and accurately impressed, perfectly finished, neatly struck by the proper organs, distinct, sharp, in due succession, and of due weight."

ENGLISH PHONETIC ELEMENTS.

The following arrangement of Bell's exhibits all the English Phonetic Elements, in a scheme of Roman letters, by means of which every detail

of English pronunciation may be exactly represented in ordinary type.

The mark (–) over vowels denotes the "long" or
name-sounds of the letters; the mark (ᴗ) denotes
their second or "short" sounds; the mark (^)
denotes the sounds of the vowel-letters before *r;*
and a dot under vowels denotes "obscure," unac-
cented sounds. The digraphs *ah, ay, aw, oo, ow,
oy,* are associated with their most usual sounds,
so as to make phonetic transcription as little as
possible different from ordinary orthography.

VOWELS.

FIRST SOUNDS.

Elements.	Illustrative words.	Elements.	Illustrative words.
1	āy	ale, day, weight.	5	ōw	old, know, beau.
2	ā	aërial, hesitate.	6	ō	obey, also.
3	ē	eel, seal, field.	7	ū	use, beauty, ague.
4	ī	idle, try, height.	8	ōo	too, through, true.

SECOND SOUNDS.

9	ă	am, carry.	12	ŏ	on, sorry.
10	ĕ	end, merit.	13	ŭ	up, hurry.
11	ĭ	ill, spirit.	14	ŏo	foot, put.

SOUNDS BEFORE R.

15	â	care, fair, there.	17	ô	ore, pour, floor.
16	ê	her, earn.	18	û	pure, cure.
	î	sir, firm.	19	ôo	poor, tour, sure.

ADDITIONAL SOUNDS.

20	ăh	ask, bath.	24	aw	wall, saw, ought.
21	ah	ah, heart, father.	25	ow	how, house, bough.
22	ahy	ay, naïve.	26	oy	boy, oil.
23	ăw	watch, want.			

OBSCURE SOUNDS.

Elements.	Illustrative words.	Elements.	Illustrative words.
27 a	a, total, collar.	80 o	-or, con-, com-.
28 e	-less, -ness, -ment.	31 u	-our, -tion, -tious.
29 i	the, -ace, -age, -ain.		

CONSONANTS.

NON-VOCAL.

82 h	hand, perhaps, vehement.	88 th	thin, bath, athwart.
83 yh	hue, human.	89 f	fine, knife, laugh.
34 wh	why, awhile.	40 p	peep, supper, hope.
85 s	say, cell, scene.	41 t	ten, matter, mate.
36 sh	wish, mission, notion.	42 k	key, cat, back, quite.
37 ch	each, fetch, church.		

VOCAL.

43 y	ye, yes, use.	52 dh	then, with, other.
44 w	we, way, beware.	53 v	vain, love, of.
45 r	ray, free, screw.	54 b	babe, rub, robber.
46 r	air, ear, ire.	55 d	did, middle, made.
47 l	let, seal, mile.	56 g	gap, gun, plague.
48 l	lure, lute, lucid.	57 m	may, blame, hammer.
49 z	zeal, as, rose.	58 n	no, tune, banner.
50 zh	vision, pleasure, rouge.	59 ng	ring, ink, uncle.
51 j	jail, jest, join.		

The letters *c*, *q*, *x*, do not appear in the above scheme, because their sounds are represented by *s* and *k*. The letter *g* appears with its "hard" sound only, because its "soft" sound is represented by *j*. The letters *ch* and *j* are retained with their ordinary associations.

Of the seven consonants denoted by digraphs, the sounds of *wh*, *th*, *sh*, *ng*, are very regularly associated with these letters; but the sounds intended by *yh, dh, zh,* are never so written in ordinary orthography.

The following tabular arrangement of English vowels will be found convenient, as showing the serial relations of the sounds.

```
pool\ 17                              1 /eel
   pull\ 16                        2 / ill
      old\ 15                    3 / ale
        ore\ 14               4 / air
          all\ 13           5 / ell
            doll\ 12       6 / an
               up\ 11     7 / a
                 err\ 10  8 / ask
                       9
                      \ah/
```

SLIDES.

In speaking, the voice is continually changing its pitch. These upward and downward movements are called *slides* or *inflections.*

Broadly speaking, there are but two inflections: the rising inflection, marked thus, (/), and the falling inflection, marked thus, (\).

The rising and falling united form what is called the falling circumflex, marked thus, (/\).

The falling and rising united form what is called the rising circumflex, marked thus, (\/).

The rising double wave consists of a falling circumflex, finished with a rising inflection, and is marked thus, (/\/).

The execution of the slides is effected by depressing the radical part of the inflection below. the middle tone for a rise, and elevating it above the middle tone for a fall.

Inflections may very truly be called the language of the emotions ; for *tones* give the *thought*, and no thought can arise in the mind without some attendant emotion; even if it be one of indifference.

What the student must know is *how* to vary his voice. The mechanism of inflections must be thoroughly mastered.

"The perfection of art is to conceal art," and the student of expression must guard continuously against allowing technique to become apparent.

The simple slides express candor, sincerity, while the compound slides are the language of artifice and double meaning.

The rising slide expresses incompleteness, anticipation, interrogation, doubt, entreaty, deference, modesty, desire ; the falling slide, completeness, satisfaction, assertion, confidence, command, disregard, haughtiness.

PICTURING.

Much of the charm of reading depends upon the reader's ability to paint the scene. The reader should locate places and objects by appropriate gestures.

Legouvé claims that the study of elocution improves the memory ; and certainly the effort necessary to retain the mental picture once made must be beneficial to the memory.

We should be careful to personate the different characters represented by suitable *looks, gestures,* and *tones.*

"Persuade yourself that there are blind men and deaf men in your audience whom you must move, interest, and persuade. Your inflection must become pantomime to the blind, and your pantomime inflection to the deaf."

Selected Readings and Recitations for Practice.

Act V., Scene II.

Baptista. Now, in good sadness, son Petruchio,
I think thou hast the veriest shrew of all.

    Petruchio. Well, I say—no: and therefore, for assurance
Let's each one send unto his wife;
And he whose wife is most obedient
To come at first when he doth send for her,
Shall win the wager which we will propose.

    Hortensio. Content.—What is the wager?

    Lucentio.   Twenty crowns.

    Pet. Twenty crowns!
I'll venture so much on my hawk or hound,
But twenty times so much upon my wife.

    Luc. A hundred, then.

    Hor.   Content.

    Pet.   A match; 'tis done.

    Hor. Who shall begin?

    Luc.   That will I.   Go,
Biondello, bid your mistress come to me.

    Biondello. I go.   [*Exit.*

    Bap. Son, I will be your half, Bianca comes.

Luc. I'll have no halves: I'll bear it all myself.

Re-enter BIONDELLO.

How now!   What news?

Bion.   Sir, my mistress sends you word
That she is busy, and she cannot come.

Pet. How! she is busy, and she cannot come!
Is that an answer?

Gremio.   Ay, and a kind one too:
Pray God, sir, your wife send you not a worse.

Pet. I hope, better.

Hor. Sirrah, Biondello, go, and entreat my wife
To come to me forthwith.   [*Exit Biondello.*

Pet.   Oho! entreat her!
Nay, then she must needs come.

Hor.   I am afraid, sir,
Do what you can, yours will not be entreated.

Re-enter BIONDELLO.

Now, where's my wife?

Bion. She says you have some goodly jest in
        hand;
She will not come; she bids you come to her.

Pet. Worse and worse; she will not come!   O
        vile,
Intolerable, not to be endured!
Sirrah, Grumio, go to your mistress;
Say I command her come to me. [*Exit Grumio.*

Hor. I know her answer.

Pet.   What?

Hor.   She will not.

Pet. The fouler fortune mine, and there an end.

Enter KATHARINA.

Bap. Now, by my holidame, here comes Katharina !

Katharina. What is your will, sir, that you send for me ?

Pet. Where is your sister, and Hortensio's wife ?

Kath. They sit conferring by the parlor fire.

Pet. Go fetch them hither; if they deny to come,

Swinge me them soundly forth unto their husbands.

Away, I say, and bring them hither straight.

[Exit Kath.

Luc. Here is a wonder, if you talk of a wonder.

Hor. And so it is; I wonder what it bodes.

Pet. Marry, peace it bodes, and love, and quiet life,

An lawful rule, and right supremacy ;

And, to be short, what not, that's sweet and happy.

Bap. Now fair befall thee, good Petruchio !

The wager thou hast won ; and I will add

Unto their losses twenty thousand crowns ;

Another dowry to another daughter,

For she is changed, as she had never been.

Pet. Nay, I will win my wager better yet ;

And show more sign of her obedience,

Her new-built virtue and obedience.

Re-enter KATH. *with* BIANCA *and* WIDOW.

See, where she comes ; and brings your froward wives

As prisoners to her womanly persuasion.—
Katharine, that cap of yours becomes you not:
Off with that bauble, throw it under foot.

> [*Kath. pulls off her cap, and throws it down.*

Widow. Lord, let me never have a cause to sigh,
Till I be brought to such a silly pass!

Bianca. Fie! What a foolish duty call you
this?

Luc. I would your duty were as foolish too:
The wisdom of your duty, fair Bianca,
Hath cost me an hundred crowns since supper-
time.

Bian. The more fool you for laying on my duty.

Pet. Katharine, I charge thee, tell these head-
strong women
What duty they do owe their lords and husbands.

Wid. Come, come, you're mocking; we will
have no telling.

Pet. Come on, I say; and first begin with her.

Wid. She shall not.

Pet. I say she shall;—and first begin with her.

Kath. Fie! fie! unknit that threat'ning unkind
brow;
And dart not scornful glances from those eyes,
To wound thy lord, thy king, thy governor;
It blots thy beauty, as frosts bite the meads;
Confounds thy fame, as whirlwinds shake fair
buds;
And in no sense is meet, or amiable.
A woman moved, is like a fountain troubled,
Muddy, ill-seeming, thick, bereft of beauty;
And, while it is so, none so dry or thirsty

Will deign to sip, or touch one drop of it.
Thy husband is thy lord, thy life, thy keeper,
Thy head, thy sovereign ; one that cares for thee,
And for thy maintenance : commits his body
To painful labor, both by sea and land ;
To watch the night in storms, the day in cold,
While thou liest warm at home, secure and safe ;
And craves no other tribute at thy hands
But love, fair looks, and true obedience ;—
Too little payment for so great a debt.
Such duty as the subject owes the prince,
Even such a woman oweth to her husband.
And when she's froward, peevish, sullen, sour,
And not obedient to his honest will,
What is she but a foul contending rebel,
And graceless traitor to her loving lord?
I am ashamed that women are so simple
To offer war, where they should kneel for peace :
Or seek for rule, supremacy, and sway,
When they are bound to serve, love, and obey.
Shakespeare.

THE MONK'S MAGNIFICAT.

A STATELY abbey many prayerful years
Had risen o'er the marshes ; thither went
In tribulation, sickness, want, or fears
The peasants for whose weal her stores were
        spent,
Certain to find a welcome and to be
Helped in the hour of their extremity.

The monks in simple ways and works were glad ;
Yet all men must have sorrows of their own.

And so a bitter grief the brothers had,
Nor mourned for other's heaviness alone.
This was the secret of their sorrowing—
That not a monk in all the house could sing!

Was it the damp air from the lonely marsh,
Or strain of scarcely intermitted prayer,
That made their voices, when they sang, as
    harsh
As any frogs' that croak in evening air—
That made less music in their hymns to lie
Than in the hoarsest wild fowl's hoarsest cry?

If love could sweeten voice to sing a song,
Theirs had been sweetest song was ever sung;
But their heart's music reached their lips all
    wrong.
The soul's intent foiled by the traitorous tongue
That marred the chapel's peace, and seemed to
    scare
The rapt devotion lingering in the air.

The brethren's prayers and fasts availing not
To give them voices sweet, their soul's desire,
The Abbot said: "Gifts He did not allot,
God at our hands will not again require;
Praise Him we must, and since we cannot praise
As we would choose, we praise Him in our
    ways."

But one good brother, anxious to remove
This, the reproach now laid on them so long,
Rejected counsel, and for very love

Besought a brother skilled in art of song
To come to them—his cloister far to leave—
And sing *Magnificat* on Christmas Eve.

And when the time for singing it had come,
With pure face raised, and sweetest voice, he
    sang:
Magnificat anima mea Dominum ; et exultavit spir-
    itus meus in Deo salutari meo.
Each in his stall the monks stood glad and
    dumb,
As through the chancel's dusk his voice out-
    rang,
Pure, clear, and perfect as the thrushes sing
Their first impulsive welcome of the spring.
At the first notes the Abbot's heart spoke low:
"O God, accept this singing, seeing we,
Had we the power, would ever praise Thee so—
Would ever, Lord, Thou know'st, sing thus for
    Thee;
Thus in our hearts Thy hymns are ever sung,
As he Thou blessest sings them with his
    tongue."

But as the voice rose higher, and more sweet,
Suscepit Israel, puerum suum, recordatus miseri-
    cordiae suae,
The Abbot's heart said: "Thou hast heard us
    grieve,
And sent an angel from beside Thy feet
To sing *Magnificat* on Christmas Eve,

To ease our ache of soul, and let us see
How we some day in heaven shall sing to
    Thee."

When, service done, the brothers gathered round
To thank the singer, modest-eyed, said he:
"Not mine the grace, if grace indeed abound,
God gave the power, if any power there be;
If I in hymn or psalm clear voice can raise,
As His the gift, so His be all the praise!"

That night—the Abbot lying on his bed—
A sudden flood of radiance on him fell,
Poured from the crucifix above his head,
And cast a stream of light across his cell—
And in the fullest fervor of the light—
An angel stood, glittering, and great, and white.
The angel spoke, his voice was low and sweet,
As the sea's murmur on low-lying shore,
Or whisper of the wind in ripened wheat.
"Brother," he said, "the God we both adore
Hath sent me down to ask, is all not right?
Why was *Magnificat* not sung to-night?"

Tranced, in the joy the angel's presence brought,
The Abbot answered: "All these weary years
We have sung our best, but always have we
    thought
Our voices were unworthy heavenly ears;
And so to-night we found a clearer tongue,
And by it the *Magnificat* was sung."

The angel answered : " All these happy years
In heaven has YOUR *Magnificat* been heard ;
This night alone the angels' listening ears
Of all its music caught no single word.
Say, who is he whose goodness is not strong
Enough to bear the burden of his song?"

The Abbot named his name. " Ah, why," he
    cried,
"Have angels heard not what we found so
    dear ?"
" Only pure hearts," the angel's voice replied,
" Can carry human songs up to God's ear ;
To-night in heaven was missed the sweetest
    praise
That ever rises from earth's mudstained maze.

" The monk who sang *Magnificat* is filled
With love of praise, and with hypocrisy ;
He sings for earth, in heaven his notes are
    stilled
By muffling weight of deadening vanity ;
His heart is chained to earth, and cannot bear
His singing higher than the listening air !

" From purest hearts most perfect music springs,
And while *you* mourned your voices were not
    sweet,
Marred by the accident of earthly things,
In heaven, God, listening, judged your song
    complete.
The sweetest of earth's music came from you,
The music of a noble life and true."

E. Nesbit.

SCENE FROM OTHELLO.

Act II., Scene III.

IAGO AND CASSIO.

Iago. What, are you hurt, lieutenant?

Cas. Ay, past all surgery.

Iago. Marry, heaven forbid!

Cas. Reputation, reputation, reputation! O, I have lost my reputation! I have lost the immortal part, sir, of myself, and what remains is bestial.—My reputation, Iago, my reputation.

Iago. As I am an honest man, I thought you had received some bodily wound; there is more offence in that, than in reputation. Reputation is an idle and most false imposition; oft got without merit, and lost without deserving. You have lost no reputation at all, unless you repute yourself such a loser. What, man! there are ways to recover the general again. You are but now cast in his mood, a punishment more in policy than in malice; even so as one would beat his offenceless dog, to affright an imperious lion. Sue to him again, and he is yours.

Cas. I will rather sue to be despised, than to deceive so good a commander, with so slight, so drunken, and so indiscreet an officer. Drunk? and speak parrot? and squabble? swagger? swear? and discourse fustian with one's own shadow?—O thou invisible spirit of wine, if thou hast no name to be known by, let us call thee—devil!

Iago. ·What was he that you followed with your sword? What had he done to you?

Cas. I know not.

Iago. Is it possible?

Cas. I remember a mass of things, but nothing distinctly; a quarrel, but nothing wherefore.—O that men should put an enemy in their mouths, to steal away their brains! that we should, with joy, revel, pleasure, and applause, transform ourselves into beasts!

Iago. Why, but you are now well enough. How came you thus recovered?

Cas. It hath pleased the devil, drunkenness, to give place to the devil, wrath; one unperfectness shows me another, to make me frankly despise myself.

Iago. Come, you are too severe a moraler. As the time, the place, and the condition of this country stands, I could heartily wish this had not befallen; but, since it is as it is, mend it for your own good.

Cas. I will ask him for my place again; he shall tell me I am a drunkard! Had I as many mouths as Hydra, such an answer would stop them all. To be now a sensible man, by and by a fool, and presently a beast! O strange! Every inordinate cup is unblessed, and the ingredient is a devil.

Iago. Come, come, good wine is a good familiar creature, if it be well used; exclaim no more against it. And, good lieutenant, I think, you think I love you.

Cas. I have well approved it, sir.—I drunk!

Iago. You, or any man living, may be drunk at some time, man. I'll tell you what you shall do. Our general's wife is now the general;—I may say so in this respect, for that he hath devoted and given up himself to the contemplation, mark, and denotement of her parts and graces :—confess yourself freely to her; importune her; she'll help to put you in your place again; she is of so free, so kind, so apt, so blessed a disposition, that she holds it a vice in her goodness, not to do more than she is requested. This broken joint between you and her husband, entreat her to splinter; and, my fortunes against any lay worth naming, this crack of your love shall grow stronger than it was before.

Cas. You advise me well.

Iago. I protest, in the sincerity of love, and honest kindness.

Cas. I think it freely; and betimes in the morning, I will beseech the virtuous Desdemona to undertake for me. I am desperate of my fortunes, if they check me here.

Iago. You are in the right. Good night, lieutenant; I must to the watch.

Cas. Good night, honest Iago.

Shakespeare.

MABEL MARTIN.

It was the pleasant harvest-time,
  When cellar-bins are closely stowed,
  And garrets bend beneath their load.

On Esek Harden's oaken floor,
  With many an autumn threshing worn,
  Lay the heaped ears of unhusked corn.

And thither came young men and maids,
  Beneath a moon that, large and low,
  Lit that sweet eve of long ago.

And jests went round, and laughs that made
  The house-dog answer with his howl,
  And kept astir the barn-yard fowl;

But still the sweetest voice was mute
  That river-valley ever heard
  From lip of maid or throat of bird;

For Mabel Martin sat apart,
  And let the hay-mow's shadow fall
  Upon the loveliest face of all.

She sat apart, as one forbid,
  Who knew that none would condescend
  To own the Witch-wife's child a friend.

The seasons scarce had gone their round,
  Since curious thousands thronged to see
  Her mother on the gallows-tree.

Few questioned of the sorrowing child,
  Or, when they saw the mother die,
  Dreamed of the daughter's agony.

Poor Mabel from her mother's grave
  Crept to her desolate hearth-stone,
  And wrestled with her fate alone.

With love, and anger, and despair,
  The phantoms of disordered sense,
  The awful doubts of Providence !

The school-boys jeered her as they passed,
  And, when she sought the house of prayer,
  Her mother's curse pursued her there.

And still o'er many a neighboring door
  She saw the horseshoe's curvéd charm,
  To guard against her mother's harm ;—

That mother, poor, and sick, and lame,
  Who daily, by the old arm-chair,
  Folded her withered hands in prayer ;—

Who turned, in Salem's dreary jail,
  Her worn old bible o'er and o'er,
  When her dim eyes could read no more !

Sore tried and pained, the poor girl kept
  Her faith, and trusted that her way,
  So dark, would somewhere meet the day.

So in the shadow Mabel sits ;
  Untouched by mirth she sees and hears,
  Her smile is sadder than her tears.

But cruel eyes have found her out,
  And cruel lips repeat her name,
  And taunt her with her mother's shame.

She answered not with railing words,
  But drew her apron o'er her face,
  And, sobbing, glided from the place.

And only pausing at the door,
　　Her sad eyes met the troubled gaze
　　Of one who, in her better days,

Had been her warm and steady friend,
　　Ere yet her mother's doom had made
　　Even Esek Harden half afraid.

He felt that mute appeal of tears,
　　And, starting, with an angry frown
　　Hushed all the wicked murmurs down.

" Good neighbors mine," he sternly said,
　　" This passes harmless mirth or jest ;
　　I brook no insult to my guest."

The broadest lands in all the town,
　　The skill to guide, the power to awe,
　　Were Harden's ; and his word was law.

None dared withstand him to his face,
　　But one sly maiden spake aside :
　　" The little witch is evil-eyed !

" Her mother only killed a cow,
Or witched a churn or dairy-pan ;
　　But she, forsooth, must charm a man !"

Poor Mabel, in her lonely home,
　　Sat by the window's narrow pane,
　　White in the moonlight's silver rain.

She strove to drown her sense of wrong,
　　And, in her old and simple way,
　　To teach her bitter heart to pray.

Poor child ! the prayer, begun in faith,
   Grew to a low, despairing cry
   Of utter misery : " Let me die !

" O God ! have mercy on Thy child,
   Whose faith in Thee grows weak and small,
   And take me ere I lose it all !"

A shadow on the moonlight fell,
   And murmuring wind and wave became
   A voice whose burden was her name.

Had then God heard her ? Had He sent
   His angel down ? In flesh and blood,
   Before her Esek Harden stood.

He laid his hand upon her arm ;
   " Dear Mabel, this no more shall be ;
   Who scoffs at you must scoff at me.

" You know rough Esek Harden well ;
   And if he seems no suitor gay,
   And if his hair is touched with gray,

" The maiden grown shall never find
   His heart less warm than when she smiled,
   Upon his knees, a little child !"

" O truest friend of all !" she said,
   " God bless you for your kindly thought,
   And make me worthy of my lot !"

He led her through his dewy fields,
   To where the swinging lanterns glowed,
   And through the doors the huskers showed.

"Good friends and neighbors!" Esek said,
  "I'm weary of this lonely life;
  In Mabel see my chosen wife!

"She greets you kindly, one and all;
  The past is past, and all offence
  Falls harmless from her innocence.

"Henceforth she stands no more alone;
  You know what Esek Harden is;—
  He brooks no wrong to him or his."

O, pleasantly the harvest moon,
  Between the shadows of the mows,
  Looked on them through the great elm-boughs.

On Mabel's curls of golden hair,
  On Esek's shaggy strength it fell;
  And the wind whispered, "It is well!"
             Whittier (Abridged).

THE KING'S DAUGHTERS.

THE King's three little daughters, 'neath the
    palace windows straying,
Had fallen into earnest talk that put an end to
    playing,
And the weary King smiled once again to hear
    what they were saying.

"It is I who love our father best!" the eldest
    daughter said;
"I am the oldest princess!" and her pretty face
    grew red.

"What is there none can do without? I love
    him more than bread!"

Then said the second princess, with her bright
    blue eyes aflame,
"Than bread? A common thing like bread!
    Thou hast not any shame!
Glad am I it is I, not thou, called by our mother's
    name.

"I love him with a *better* love than one so tame
    as *thine ;*
More than—oh, what then shall I say that is both
    bright and fine,
And is not common? Yes, I know—I love him
    more than wine!"

Then the little youngest daughter, whose speech
    would sometimes halt
For her dreamy way of thinking, said, "You are
    both in fault ;
'Tis I who love our father best—I love him more
    than salt!"

Shrill little shrieks of laughter greeted her latest
    words,
And the two joined hands, exclaiming, "But this
    is most absurd!"
And the King, no longer smiling, was grieved that
    he had heard,

For the little youngest daughter, with her eyes
    of steadfast gray,

Could always move his tenderness, and charm his
    care away.
"She grows more like her mother dead," he
    whispered, "day by day.

"But she is very little, and I will find no fault
That while her sisters strive to see who most shall
    me exalt,
She holds me nothing dearer than a common
    thing like salt."

The portly cook was standing in the courtyard by
    the spring;
He winked and nodded to himself: "That little
    quiet thing
Knows more than both the others, as I shall show
    the King."

That afternoon at dinner there was nothing fit to
    eat;
The King turned, frowning angrily, from soup and
    fish and meat,
And he found a cloying sweetness in the dishes
    that were sweet.

"And yet," he muttered, musing, "I cannot find
    the fault;
Not a thing has tasted like itself but this honest
    cup of malt."
Said the youngest princess, shyly, "Dear father,
    they want salt."

A sudden look of tenderness shone on the King's
    dark face,

As he sat his little daughter in the dead Queen's
    vacant place ;
And he thought, "She has her mother's heart—
    aye,. and her mother's grace.

" Great love through smallest channels will find
    its surest way ;
It waits not state occasions, which may not come,
    or may ;
It comforts and it blesses hour by hour and day
    by' day."
    Margaret Vandergrift (*From the German*).

THE JUDGMENT OF ST. PANCRATIUS.

GREAT Diocletian in his judgment court
    Appeared, by all his pomp of majesty
Compassed and guarded ; lion-like his port;
    Then whispered man to man: "That terrible
      eye
Without yon lictors' axes or their rods
Will drive the renegade to his country's gods."

Pancratius entered—entered with a smile ;
    Bowed to the Emperor ; next to those around,
First east, then west. The Emperor gazed awhile
    On that bright countenance ; knew its import ;
      frowned :
"A malefactor known ! Yet there you stand !
Young boy, be wise in time. Hold forth your hand !

"Yon censer mark ! It comes from Jove's chief
    fane ;
    See next yon vase cinctured with flower-attire :

Lift from that vase its smallest incense-grain;
  Commit it softly to yon censer's fire:
Your father, boy, was well with me; and I
Would rather serve his son than bid him die."

Pancratius mused a moment, then began:
  "Emperor, 'tis true I am a boy; no more:
But one within me changes boy to man,
  Christ—God and Man : that Lord the just adore.
A pictured lion hangs above thy head:
Say, can a picture touch man's heart with dread?

"Thou, too, great Emperor, art but pictured life:
  He only lives who quickened life in all:
Men are but shadows: in a futile strife
  They chase each other on a sun-bright wall.
Shadows are they the hosts that round thee throng;
Shadows their swords that vindicate this wrong.

"What gods are those thou bidst me serve and
      praise?
  Adulterers, murderers, gods of fraud and theft.
If slave of thine walked faithful in their ways
  What were his sentence? Eyes of light bereft;
The scourge, the rope! Our God is good. His
    name
Paints on His servant's face no flush of shame.

"Exteriorly, 'tis true, thy gods are great,
  They and their sort: this hour they rule the
      lands:
Ay, but, expectant at an unbarred gate,
  A greatness of a different order stands,—

The Babe of Bethlehem's. He thy gods shall slay,
Though small His hand, and rend earth's chain
    away."

The Emperor shook: as one demon-possessed
    He glared upon that youth; his wan cheek
        burned :
With wonder dumb panted his struggling breast :
    Silent to that prætorian guard he turned—
He pointed to Pancratius.   "Let him die!"
Pancratius stood, and pointed to the sky.
                              Aubrey de Vere.

THE GOLDEN BRIDGE.

LET him listen, whoso would know,
Concerning the wisdom of King Tee Poh.

Fair is Pekin, with round it rolled
Wave on wave of its river of gold ;
They gird its walls with their ninefold twine,
And the bridges that cross them are ninety and
    nine.
And as soon as the wind of morning blows,
And the gray in the East takes a fleck of rose,
Upon each bridge 'gins the shuffle and beat
Of hundreds of hoofs and thousands of feet ;
And all day long there is dust and dinne
And the coolie elbows the mandarin,
And gibe is given and oath and blow—
'Twas thus in the time of King Tee Poh.

It grieved the King that it should be so ;
Then out of his wisdom spoke King Tee Poh :

" Build me a hundredth bridge, the best,
Higher and wider than all the rest,
With posts of teak and cedarn rails
And planks of sandal, with silver nails;
Gild it and paint it vermilion red, ·
And over it place the dragon's head;
And be it proclaimed to high and low
That over this fortunate arch shall go
Passenger none that doth not throw
Golden toll to the river below.
And when the piece of gold is cast,
Thrice let the trumpets sound a blast :
And the mandarin write with respectful look
The passenger's name in a silken book,
So that I, the King, may have in hand
The list of the wealthiest of my land."

 Straightway the bridge was builded so
 As had spoken the wisdom of King Tee Poh.

And every day, from dawn till dark,
They who watched the fortunate arch could
  mark
Like a cloud of midges that glow and gleam,
The gold toll cast to the hurrying stream ;
And all day the trumpet sounded loud,
And the mandarin of the guard kowtowed,
As he wrote the name, with respectful look,
Of the passenger high in his silken book ;
And all the while grew the renown
Of the fortunate arch in Pekin town,
Till of the wealthiest it was told,
" He spends his day on the bridge of gold."

And when a month and a day were spent,
The King Tee Poh for his treasurer sent
" Go to the bridge," said he, " and look
At the list of names in the silken book,
And of all that are written, small and great,
Confiscate to me the estate ;
As the sage Confucius well doth show,
A wealthy fool is the State's worst foe."

And the treasurer whispered, bending low,
" Great is the wisdom of King Tee Poh."

George T. Lanigan, in " The Century."

THE BALLAD OF THE STRANGER.

'T was off the coast of Scarboro'
  In sixteen eighty-three ;
An April night fell lowering
  Upon an angry sea.
And on the heights above the town
Was many a watcher gazing down,
And murmuring with a shrug and frown :
  " A woeful night 't will be !"

The wind across the surges
  Came howling to the land ;
In foaming wrath the breakers
  Came bounding on the strand ;
When with a voice from turret high
Sounded aloud that startled cry :
" A wreck ! a wreck !—Shoremen ahoy !
  She's plunging for the land !"

Down from the heights went skurrying
   The wreckers to the shore,
And women wild, who seaward smiled
   Hopeful an hour before!
The ship—Great God!—in flames her prow!—
The flames are bursting from her bow!
She speeds full sail!—
                Thank Heaven the gale
   Is blowing to the shore!

Red are the waves before her—
   Each crest a flaming brand!
With tongues of wrath and fiery breath
   She leaps toward the strand.
"Ahoy! ahoy!"—the trumpet rings—
See! on the hidden reef she springs!
To rock she clings,—
                On rock she swings
   Her larboard to the land.

A thousand shrieks of terror
   Arise from ship and shore!
"Launch! launch the boats!"—the trumpet
     notes
, Blare out above the roar.
But every boat, from beach or deck,
Like shells the breakers crush and wreck.
Stranded she stood . . .
              In fire and flood . . .
   But a hundred yards from shore.

Down to the beach a stranger
   Stept calmly thro' the crowd;

He doffed his cloak, and up he spoke
  With startling voice and loud :
" Come on with me, the bravest three ! " . . .
(In yawl they plunged into the sea.)
" Give me the rope !—
                Cowards are we,
  To cringe at watery shroud ?"

Athwart the breakers plunging
  Went gallant men and yawl ;
A rope they bore, the coil on shore
  Trailed out with snaky crawl.
See ! heavens ! they sink !—
             A mountain wave
Buries them deep in yawning grave !
A shriek ! a wail from women pale
  The bravest souls appall.

Up ! see !—the dauntless heroes
  Upon the surges rise !
" Praise God !" a shout from ship and shore
  Breaks upwards to the skies.
" Courage !"—peals out that stranger's shout,—
He strikes the wreck . . .
            He leaps on deck . . .
His rope ties fast to mizzen mast,
  And, " *Down the rope !*" he cries.

Swift, one by one, like pigeons
  From startled cote, they pour—
They glide on rope through breakers
  Hand over hand to shore . . .

The flames! the flames!

      With hiss and gnash
Sternward their tongues of fire they flash,
And on the flames the surges dash
 With seething shriek and roar!

The last man's o'er the taffrail—
 Alone the stranger . . . No!
Horrors!—up from the hatchway
 A woman from below!—
Clasping her child, in terror wild
Shrieking:
    "O God! my child! my child!"
To the stranger's breast her babe she prest
 In agony of woe.

Tho' singed with fire, that hero
 To his breast the babe he bound;
Then to the sea leapt mother and he—
 She clasping him around.
Now on the rope, hand over hand,
Thro' breakers plunging for the strand—
"Hold to the rope! it *burns!*"
      From land
 Rings out the trumpet-sound.

A shuddering cry uprises
 From thousands on the lee—
The rope it parts, and flaming darts
 And hisses in the sea!
"Hold to the rope!"
     Alas! a wave
O'erwhelms him deep—that hero brave!

Down, down they sink into that grave—
  The mother, babe, and he.

There is a sudden silence
  Hushes the land in awe,
As over the sands an hundred hands
  That willing rope they draw. . . .
" PRAISE GOD, THE LORD !"
                       Bursts sudden cry
From thousand voices raised on high. . . .
See ! on the land, above the strand,
Silent and pale they lie !

In fixéd grasp that hero
  The rope still firmly holds !
And firm his teeth with clench of death
  That mother's sleeve enfolds !
Oh, fearful sight !—more ghastly seem
Those faces in the lurid gleam. . . .
But—hark ! he speaks !
                  He stirs ! he wakes !
He starts as from a dream !

And the mother's lips are quivering
  As if to speak . . . and hark !
She calls her child . . . she gazes wild
  Toward the burning barque.
The stranger smiled; unbound his breast . . .
The babe lay smiling in its nest !
The mother shrieked in rapture wild :
" My child ! my child !—
                Thank God ! my child !"

The multitude came surging,
  And round that stranger prest,—
With prayer and cry that reached the sky
  That hero brave they blest.
But not a word the stranger spoke . . .
He calmly smiled,—
           He donned his cloak,
And, bowing, vanished in the dark.
" *Who was the hero ?*" . . . EBEN STARK !

BRIDGE OF SIGHS.

ONE more Unfortunate,
  Weary of breath,
Rashly importunate,
  Gone to her death !

Take her up tenderly,
  Lift her with care ;—
Fashioned so slenderly,
  Young, and so fair !

Look at her garments,
Clinging like cerements ;
  Whilst the wave constantly
Drips from her clothing ;
  Take her up instantly,
Loving, not loathing.

Touch her not scornfully,
Think of her mournfully,
  Gently and humanly ;
Not of the stains of her,

All that remains of her
  Now, is pure womanly.

Make no deep scrutiny
Into her mutiny
  Rash and undutiful ;
Past all dishonor,
Death has left on her
  Only the beautiful.

Still, for all slips of hers,
  One of Eve's family—
Wipe those poor lips of hers,
  Oozing so clammily ;

Loop up her tresses
  Escaped from the comb,
Her fair auburn tresses ;
Whilst wonderment guesses
  Where was her home ?

Who was her father ?
Who was her mother ?
Had she a sister ?
Had she a brother ?
  Or was there a dearer one
  Still, and a nearer one
Yet, than all other ?

Alas ! for the rarity
Of Christian charity
  Under the sun !
Oh ! it was pitiful !

Near a whole city full,
  Home she had none.

Sisterly, brotherly,
Fatherly, motherly,
  Feelings were changed ;
Love, by harsh evidence,
Thrown from its eminence
Even God's providence
  Seeming estranged.

Where the lamps quiver
So far in the river,
  With many a light
From window and casement,
From garret to basement,
She stood, with amazement,
  Houseless by night.

The bleak winds of March
  Made her tremble and shiver ;
But not the dark arch,
  Or the black flowing river :
Mad from life's history,
Glad to death's mystery
  Swift to be hurled—
Anywhere, anywhere
  Out of the world!

In she plunged boldly,
No matter how coldly
  The rough river ran,—
Over the brink of it,

Picture it,—think of it,
  Dissolute man!
Lave in it, drink of it
  Then, if you can!

Take her up tenderly,
  Lift her with care;
Fashioned so slenderly,
  Young, and so fair!
Ere her limbs frigidly
Stiffen so rigidly,
  Decently,—kindly,—
Smooth and compose them;
And her eyes, close them,
  Staring so blindly!

Dreadfully staring
  Through muddy impurity,
As when with the daring
Last look of despairing
  Fixed on futurity.

Perishing gloomily,
Spurred by contumely,
Cold inhumanity,
Burning insanity,
  Into her rest,—
Cross her hands humbly,
As if praying dumbly,
  Over her breast!

Owning her weakness,
　Her evil behavior,
And leaving, with meekness,
　Her sins to her Saviour!

Hood.

DE PROFUNDIS.

THE face, which, duly as the sun,
Rose up for me with life begun,
To mark all bright hours of the day
With hourly love, is dimmed away,—
And yet my days go on, go on.

The tongue, which, like a stream, could run
Smooth music from the roughest stone,
And every morning with "Good-day"
Made each day good, is hushed away,—
And yet my days go on, go on.

The heart, which, like a staff, was one
For mine to lean and rest upon,
The strongest on the longest day
With steadfast love, is caught away,—
And yet my days go on, go on.

And cold before my summer's done,
And deaf in Nature's general tune,
And fallen too low for special fear,
And here, with hope no longer here,—
While the tears drop, my days go on.

The world goes whispering to its own,
"This anguish pierces to the bone;"

And tender friends go sighing round,
"What love can ever cure this wound?"
My days go on, my days go on.

The past rolls forward on the sun,
And makes all night.   O dreams begun,
Not to be ended!   Ended bliss,
And life that will not end in this!—
My days go on, my days go on.

Breath freezes on my lips to moan :
As one alone, once not alone,
I sit and knock at Nature's door,
Heart-bare, heart-hungry, very poor,
Whose desolated days go on.

I knock and cry, Undone, undone!
Is there no help, no comfort,—none?
No gleaning in the wide wheat-plains
Where others drive their loaded wains?—
My vacant days go on, go on.

This Nature, though the snows be down,
Thinks kindly of the bird of June :
The little red hip on the tree
Is ripe for such.   What is for me
Whose days so winterly go on?

No bird am I to sing in June,
And dare not ask an equal boon.
Good nests and berries red are Nature's
To give away to better creatures,—
And yet my days go on, go on.

I ask less kindness to be done,—
Only to loose these pilgrim-shoon,
(Too early worn and grimed) with sweet
Cool deathly touch to these tired feet,
Till days go out which now go on.

Only to lift the turf unmown
From off the earth where it has grown,
Some cubit-space, and say, "Behold!
Creep in, poor heart, beneath that fold,
Forgetting how the days go on."

What harm would that do?   Green anon
The sward would quicken, overshone
By skies as blue ; and crickets might
Have leave to chirp there day and night,
While my new rest went on, went on.

From gracious Nature have I won
Such liberal bounty ? may I run
So, lizard-like, within her side,
And there be safe, who now am tried
By days that painfully go on ?

—A Voice reproves me thereupon,
More sweet than Nature's when the drone
Of bees is sweetest, and more deep
Than when the rivers overleap
The shuddering pines, and thunder on.

God's voice, not Nature's!   Night and noon
He sits upon the great white throne,
And listens for the creatures' praise.

What babble we of days and days ?
The Dayspring He, whose days go on.

He reigns above, He reigns alone ;
Systems burn out, and leave His throne ;
Fair mists of seraphs melt and fall
Around Him, changeless amid all,—
Ancient of days, whose days go on.

He reigns below, He reigns alone,
And, having life in love foregone
Beneath the crown of sovran thorns
He reigns the jealous God.   Who mourns
Or rules with Him, while days go on ?

By anguish which made pale the sun,
I hear Him charge His saints that none
Among His creatures anywhere
Blaspheme against Him with despair,
However darkly days go on.

Take from my head the thorn-wreath brown !
No mortal grief deserves that crown.
O sùpreme love, chief misery,
The sharp regalia are for *Thee*
Whose days eternally go on !

For us, whatever's undergone,
Thou knowest, willest, what is done.
Grief may be joy misunderstood :
Only the Good discerns the good.
I trust Thee while my days go on.

Whatever's lost, it first was won ;
We will not struggle nor impugn.
Perhaps the cup was broken here,
That heaven's new wine might show more clear.
I praise Thee while my days go on.

I praise Thee while my days go on ;
I love Thee while my days go on ;
Through dark and dearth, through fire and
    frost,
With emptied arms and treasure lost,
I thank Thee while my days go on.

And having in Thy life-depth thrown
Being and suffering (which are one),
As a child drops his pebble small
Down some deep well, and hears it fall
Smiling,—so I. *Thy days go on.*

                          Mrs. Browning.

THE FACE AGAINST THE PANE.

MABEL, little Mabel,
  With face against the pane,
Looks out across the night
And sees the Beacon Light
  A-trembling in the rain.
She hears the sea-birds screech,
And the breakers on the beach
  Making moan, making moan.
And the wind about the eaves
Of the cottage sobs and grieves ;

And the willow-tree is blown
   To and fro, to and fro,
Till it seems like some old crone
Standing out there all alone,
   With her woe,
Wringing, as she stands,
Her gaunt and palsied hands !
While Mabel, timid Mabel,
   With face against the pane,
Looks out across the night
And sees the Beacon Light
   A-trembling in the rain.

Set the table, maiden Mabel,
   And make the cabin warm ;
Your little fisher-lover
   Is out there in the storm,
And your father—you are weeping !
   O Mabel, timid Mabel,
   Go, spread the supper-table,
And set the tea a-steeping.
Your lover's heart is brave,
   His boat is staunch and tight :
And your father knows the perilous reef
   That makes the water white.
—But Mabel, darling Mabel,
   With face against the pane,
Looks out across the night
   At the Beacon in the rain.

The heavens are veined with fire !
   And the thunder, how it rolls !

In the lullings of the storm
   The solemn church-bell tolls
   For lost souls !
But no sexton sounds the knell
   In that belfry old and high ;
Unseen fingers sway the bell
   As the wind goes tearing by !
How it tolls for the souls
   Of the sailors on the sea !
God pity them, God pity them,
   Wherever they may be !
God pity wives and sweethearts
   Who wait and wait in vain !
And pity little Mabel,
   With face against the pane.

A boom !—the Light-house gun !
   (How its echo rolls and rolls !)
'Tis to warn the home-bound ships
   Off the shoals !
See ! a rocket cleaves the sky
   From the Fort,—a shaft of light !
See ! it fades, and, fading, leaves
   Golden furrows on the night !
What made Mabel's cheek so pale ?
   What made Mabel's lips so white ?
Did she see the helpless sail
   That, tossing here and there,
   Like a feather in the air,
Went down and out of sight ?
Down, down, and out of sight !
Oh, watch no more, no more,

With face against the pane ;
  You cannot see the men that drown
    By the Beacon in the rain!

From a shoal of richest rubies
  Breaks the morning clear and cold ;.
And the angel on the village spire,
  Frost-touched, is bright as gold.
Four ancient fishermen,
  In the pleasant autumn air,
Come toiling up the sands,
With something in their hands,—
Two bodies stark and white,
Ah, so ghastly in the light,
  With sea-weed in their hair !

O ancient fishermen,
  Go up to yonder cot !
You'll find a little child,
  With face against the pane,
Who looks toward the beach,
  And, looking, sees it not.
She will never watch again !
  Never watch and weep at night !
For those pretty, saintly eyes
Look beyond the stormy skies,
  And they see the Beacon Light.
                    T. B. Aldrich.

THE GAME KNUT PLAYED.

A PAGE who seemed of low degree,
And bore the name of Knut, was he ;
The high-born Princess Hilga she.

And that the youth had served her long,
Being quick at errands, skilled in song,
To jest with him she thought no wrong.

And so it chanced one summer day,
At chess, to while the time away,
The page and princess sat at play.

At length she said, "To play for naught
Is only sport to labor brought,
So let a wager guerdon thought."

He answered, "Lady, naught have I
Whose worth might tempt a princess high
Her uttermost of skill to try."

"And yet this ruby ring," she said,
"I'll risk against the bonnet red
With snow-white plume that crowns thy head.

"And should I win, do not forget,
Or should I lose, whichever yet,
I'll take my due, or pay my debt."

And so they played, as sank the sun ;
But when the game they played was done,
The page's cap the princess won.

"My diamond necklace," then she cried,
"I'll match against thy greatest pride,
The brand held pendent at thy side."

" Not so," he said—" that tempered glaive,
Borne oft by noble hands and brave,
To me my dying father gave.

" Fit only for a true man's touch,
I hold it dear and prize it much—
No diamond necklace mates with such.

" But, though my father's ghost be wroth,
I'll risk the weapon, nothing loth,
Against thy love and virgin troth."

Reddened her cheeks at this in ire,
This daughter of a royal sire,
And flashed those eyes of hers like fire.

" Thy words, bold youth, shall work thee ill :
Thou canst not win against my skill,
But I can punish at my will.

" Begin the game ; that hilt so fine
Shall nevermore kiss hand of thine,
Nor thou again be page of mine !"

Answered the page : " Do not forget,
Or win or lose, whichever yet,
I'll take my due, or pay my debt.

" And let this truth the end record :
I risk to-day my father's sword
To be no more thy page, but lord."

Down sat the pair to play once more,
Hope in his bosom brimming o'er,
And hers with pride and anger sore.

From square to square the bishops crept,
The agile knights eccentric leapt,
The castles onward stately swept.

Pawns fell in combat one by one ;
Knights, rooks, and bishops could not shun
Their fate before that game was done.

Well fought the battle was, I ween,
Until two castles and a queen
Guarding the kings alone were seen.

"Check !" cried the princess, all elate ;
"Check !" cried the page, and sealed the fate
Of her beleaguered king with " mate !"

The princess smiled, and said : " I lose,
Nor can I well to pay refuse—
From my possessions pick and choose.

" Or diamonds bright, or chests of gold,
Or strings of pearls of worth untold,
These may be thine to have and hold :

" Or costly robes to feed thy pride,
Or coursers such as monarchs ride,
Or castles tall, or manors wide—

" Any or all of such be thine ;
But, save he spring from royal line,
No husband ever can be mine."

" Nor jewels rich, nor lands in fee,
Steeds, robes, nor castles pleasure me ;
Thy love and troth be mine," said he.

" Nor shalt thou lack of state and pride
When seated crowned thy lord beside,
As Knut, the King of Denmark's bride !"

Ring marriage-bells from sun to sun,
And tell the gossips, as they run,
How Sweden's princess has been won.

THE ROMANCE OF THE SWAN'S NEST.

" So the dreams depart,
So the fading phantoms flee,
And the sharp reality
Now must act its part."
Westwood's " Beads from a Rosary."

LITTLE Ellie sits alone
'Mid the beeches of a meadow,
By a stream-side on the grass,
And the trees are showering down
Doubles of their leaves in shadow,
On her shining hair and face.

She has thrown her bonnet by,
And her feet she has been dipping
In the shallow water's flow ;

Now she holds them nakedly
  In her hands, all sleek and dripping,
    While she rocketh to and fro.

Little Ellie sits alone,
  And the smile she softly uses
    Fills the silence like a speech,
While she thinks what shall be done,
  And the sweetest pleasure chooses
    For her future within reach.

Little Ellie in her smile
  Chooses, " I will have a lover,
    Riding on a steed of steeds.
He shall love me without guile,
  And to *him* I will discover
    The swan's nest among the reeds.

" And the steed shall be red-roan,
  And the lover shall be noble,
    With an eye that takes the breath.
And the lute he plays upon
  Shall strike ladies into trouble,
    As his sword strikes men to death.

" And the steed it shall be shod
  All in silver, housed in azure ;
    And the mane shall swim the wind ;
And the hoofs along the sod
  Shall flash onward, and keep measure,
    Till the shepherds look behind.

"But my lover will not prize
  All the glory that he rides in,
    When he gazes in my face.
He will say, ' O Love, thine eyes
  Build the shrine my soul abides in,
    And I kneel here for thy grace!'

"Then, ay, then he shall kneel low,
  With the red-roan steed anear him,
    Which shall seem to understand,
Till I answer, ' Rise and go!'
  For the world must love and fear him
    Whom I gift with heart and hand.

"Then he will arise so pale,
  I shall feel my own lips tremble
    With a *yes* I must not say :
Nathless maiden-brave, ' Farewell,'
  I will utter, and dissemble—
    ' Light to-morrow with to-day!'

"Then he'll ride among the hills
  To the wide world past the river,
    There to put away all wrong,
To make straight distorted wills,
  And to empty the broad quiver
    Which the wicked bear along.

"Three times shall a young foot-page
  Swim the stream, and climb the mountain,
    And kneel down beside my feet:
' Lo! my master sends this gage,

Lady, for thy pity's counting.
　　What wilt thou exchange for it ?'

" And the first time I will send
　A white rosebud for a guerdon :
　　And the second time, a glove ;
But the third time I may bend
　From my pride, and answer,—' Pardon,
　　If he comes to take my love.'

" Then the young foot-page will run ;
　Then my lover will ride faster,
　　Till he kneeleth at my knee :
' I am a duke's eldest son,
　Thousand serfs do call me master,
　　But, O Love, I love but *thee* l'

" He will kiss me on the mouth
　Then, and lead me as a lover
　　Through the crowds that praise his
　　　deeds.
And, when soul-tied by one troth,
　Unto *him* I will discover
　　That swan's nest among the reeds."

Little Ellie, with her smile
　Not yet ended, rose up gayly,
　　Tied the bonnet, donned the shoe,
And went homeward, round a mile,
　Just to see, as she did daily,
　　What more eggs were with the two.

Pushing through the elm-tree copse,
 Winding up the stream, light-hearted,
  Where the osier pathway leads,
Past the boughs she stoops, and stops.
 Lo, the wild swan had deserted,
  And a rat had gnawed the reeds!

Ellie went home sad and slow.
 If she found the lover ever,
  With his red-roan steed of steeds,
Sooth I know not; but I know
 She could never show him—never,
  That swan's nest among the reeds.

      Mrs. Browning.

A MESSAGE.

It was Spring in the great city—every gaunt and
 withered tree
Felt the shaping and the stir at heart of leafy
 prophecy;
All the wide-spread umber branches took a tender tint of green,
And the chattering brown-backed sparrow lost
 his pert, pugnacious mien
In a dream of mate and nestlings shaded by a
 verdant screen.

It was Spring—the grim ailanthus, with its snaky
 arms awry,
Held out meagre tufts and bunches to the sun's
 persistency:

Every little square of greensward, railed in from
    the dusty way,
Sent its straggling forces upward, blade and spear
    in bright array,
While the migratory organs Offenbach and
    Handel play.

Through the heart of the vast Babel, where the
    tides of being pour,
From his labor in the evening came the sturdy
    stevedore,
Towering like a son of Anak, of a coarse, ungainly
    mould ;
Yet the hands begrimed and blackened in the
    hardened fingers hold
A dandelion blossom, shining like a disk of gold.

Wayside flower ! with thy plucking did remem-
    brance gently lay
Her hand upon the tomb of youth and roll the
    stone away ?
Did he see a barefoot urchin wander singing up
    the lane,
Carving from the pliant willow whistles to pro-
    long the strain,
While the browsing cows, slow driven, chime
    their bells in low refrain ?

Did his home rise up before him, and his child,
    all loving glee,
Hands and arms in eager motion for the golden
    mystery ?

Or the fragile, pallid mother, seeing in that starry
    eye
God's eternal, fadeless garden, God's wide sun-
    shine and His sky,
Hers through painless, endless ages, bright'ning
    through immensity?

None may know—the busy workings of the brain
    remain untold,
But the loving deed—the outgrowth—brings us
    lessons manifold.
Smiles and frowns—a look—a flower growing by
    the common way,
Trifles born with every hour make the sum of
    life's poor day,
And the jewels that we garner are the tears we
    wipe away.

Scribner's Monthly.

WASHINGTON.

IT matters very little what spot may have been
the birthplace of Washington. No people can
claim, no country can appropriate him. The
boon of Providence to the human race, his fame
is eternity, and his residence creation. Though
it was the defeat of our arms, and the disgrace of
our policy, I almost bless the convulsion in which
he had his origin. If the heavens thundered, and
the earth rocked, yet, when the storm had passed,
how pure was the atmosphere that it cleared!
How bright, on the brow of the firmament, was
the planet which it revealed to us!

In the production of Washington, it seems as if Nature was endeavoring to improve upon herself, and that all the virtues of the ancient world were but so many studies preparatory to the patriot of the new. Individual instances, no doubt, there were, splendid exemplifications of some singular qualification. Cæsar was merciful, Scipio was temperate, Hannibal was patient; but it was reserved for Washington to blend them all in one, and like the lovely masterpiece of the Grecian artist, to exhibit, in one glow of associated beauty, the pride of every model, and the perfection of every master.

As a general he marshalled the farmer into a veteran, and supplied by discipline the absence of experience. As a statesman he enlarged the policy of the cabinet into the most comprehensive system of general advantage; and such was the wisdom of his views, and the philosophy of his counsels, that to the soldier and statesman he almost added the character of the sage! A conqueror, he was untainted with the crime of blood; a revolutionist, he was free from any stain of treason, for aggression commenced the contest, and his country called him to command. Liberty unsheathed his sword, necessity stained, victory returned it.

If he had paused here, history might have doubted what station to assign him; whether at the head of her citizens or her soldiers, her heroes or her patriots. But the last glorious act crowns his career, and banishes all hesitation. Who like

Washington, after having emancipated a hemi-
sphere, resigned its crown, and preferred the
retirement of domestic life to the adoration of a
land he might almost be said to have created?
Happy, proud America! The lightnings of
heaven yielded to her philosophy. The tempta-
tions of earth could not seduce her patriotism.

Charles Phillips.

" PERSEVERE."

ROBERT, the Bruce, in the dungeon stood
  Waiting the hour of doom;
Behind him the Palace of Holyrood,
  Before him, a nameless tomb.
And the foam on his lip was flecked with red,
As away to the past his memory sped,
Upcalling the day of his great renown
When he won and he wore the Scottish crown;
  Yet come there shadow, or come there shine,
  The spider is spinning his thread so fine.

"I have sat on the royal seat of Scone,"
  He muttered, below his breath;
"It's a luckless change, from a kingly throne
  To a felon's shameful death."
And he clinched his hand in his despair,
And he struck at the shapes that were gathering
      there
Pacing his cell in impatient rage,
As a new-caught lion paces his cage.

"Oh, were it my fate to yield up my life
  At the head of my liegemen all !
In the foremost shock of the battle-strife
  Breaking my country's thrall,
I'd welcome death from the foeman's steel,
Breathing a prayer for old Scotland's weal ;
But here, where no pitying heart is nigh,
By a loathsome hand, it is hard to die.

"Time and again have I fronted the pride
  Of the tyrant's vast array.
But only to see, on the crimson tide,
  My hopes swept far away.
Now a landless chief, and a crownless king,
On the broad, broad earth not a living thing
To keep me court, save yon insect small
Striving to reach from wall to wall.

"Work—work as a fool, as I have done,
  To the loss of your time and pain—
The space is too wide to be bridged across,
  You but waste your strength in vain."
And Bruce for the moment forgot his grief,
His soul now filled with the same belief,
That howsoever the issue went,
For evil or good was the omen sent.

As a gambler watches his turning card
  On which his all is staked ;
As a mother waits for the hopeful word
  For which her soul has ached ;

It was thus Bruce watched, with every sense
Centred alone in that look intense ;
All rigid he stood with unuttered breath,
Now white, now red, but still as death.

Six several times the creature tried,
　When at the seventh : " See—see !
He has spanned it over," the captive cried—
　" Lo ! a bridge of hope to me ;
Thee, God, I thank—for this lesson here
Has tutored my soul to Persevere !"
And it served him well, for ere long he wore
In freedom the Scottish crown once more ;
　And come there shadow, or come there shine,
　The spider is spinning his thread so fine.

John Brougham.

THE CHAMBERED NAUTILUS.

THIS is the ship of pearl which, poets feign,
Sails the unshadowed main :
The venturous bark that flings
On the sweet summer wind its purple wings
In gulfs enchanted, where the siren sings,
And coral reefs lie bare ;
Where the cold sea-maids rise to sun their
　　streaming hair.

Its webs of living gauze no more unfurl ;
Wrecked is the ship of pearl !
And every chambered cell
Where its dim-dreaming life was wont to dwell,
As the frail tenant shaped his growing shell,

Before thee lies revealed :
Its irised ceiling rent, its sunless crypt unsealed.

Year after year beheld the silent toil
That spread the lustrous coil :
Still as the spiral grew,
He left the past year's dwelling for the new,
Stole with soft step its shining doorway through,
Built up its idle door,
Stretched in his last-found home, and knew the
     old no more.

Thanks for the heavenly message brought by
     thee,
Child of the wandering sea,
Cast from her lap, forlorn !
From thy dead lips a clearer note is born
Than ever Triton blew from wreathèd horn !
While on mine ear it rings,
Through the deep caves of thought I hear a voice
     that sings :

Build thee more stately mansions, O my soul,
As the swift seasons roll !
Leave thy low-vaulted past !
Let each new temple, nobler than the last,
Shut thee from heaven with a dome more vast,
Till thou at length art free,
Leaving thine outgrown shell by life's unresting
     sea.

                        Oliver W. Holmes.

CHARONDAS.

HE lifted his forehead and stood at his height,
And gathered his cloak round his noble age,
   This man, the law-giver, Charondas the Greek :
   And loud the Eubœans called to him : " Speak,
We listen and learn, O Sage !"

" In peace shall ye come where the people be,"
Spake the lofty figure with flashing eyes :
   " But whoso comes armed to the public hall
   Shall suffer his death before us all."
And the hearers believed him wise.

The years sped quick, and the years dragged
      slow :
In council oft was the throne arrayed.
   But never the statued chamber saw
   The gleam of weapon ; for loving law
The Greeks from their hearts obeyed.

War's challenge knocked at the city gates :
Students flocked to the front, grown bold :
   The strong men, girded, faced up to the North :
   The women wept to the gods ; and forth
Went the brave of the days of old.

Peace winged her flight to the city gates :
Young men and strong, they followed fast
   Back to the breast of their fair, free land :
   Charondas, afar on the foreign strand,
Remained at his post the last.

Their leader he, in war as in word,
The fire of youth for his life-long lease,
  The strength of Mars in the arm that stood
  Seven hot decades upheld for good
In the turbulent courts of Greece.

The fight is finished, the council meets.
Who is the tardy comer without,
  In cuirass and shield, and with clanking sword,
  Who strides up the aisles without a word,
Rousing that awe-struck shout?

The tardy comer, home from the field,
Great gods! the first to forget and belie
  The law he honored, the law he formed:
  " Charondas, stand! *You enter armed*,"
With a shudder the hundreds cry.

The men who loved him on every side,
The men he led to the victors' gain.
  He paused a moment, the fearless Greek,
  A sudden glow on his ashen cheek,
A sudden thought in his brain.

" I seal the law with my soul and might:
I do not break it," Charondas said.
  He raised his blade, and plunged to the hilt;
  Ah! vain they rush, for in glory and guilt,
He lay on the marble, dead.

                Louise Imogen Guiney.

AMERICA'S DEBT TO FRANCE.

IT may perhaps be suggested that the fact that
France lavished her favors on the American people in the past does not explain her present action. Logically—the objector may say—America
should send bronze statues to France, not France
to America. We never sent armed men to her
aid when all Europe was banded against her.
While her land was overrun, and German, Rus-
sian, English armies swept over her fields and
towns, leaving a track of ruin behind them, only
French blood was shed in her behalf. Our ships
did not go down with French ships at Trafalgar,
our treasure did not melt away in the fiery fur-
nace of French tribulation and German triumph.
If we are paying taxes to support our credit and
diminish our debt, no part of that debt was incurred to save French interests or French terri-
tory. True—but he knows little of the hidden
springs that control human action who does not
know that there is no gratitude like that which is
felt by the benefactor. It is far easier to forget
the favors that we have received than those that
we have conferred. That pattern of shrewd
worldly wisdom, Benjamin Franklin, ingenuously
tells us that when he wanted to secure the good-
will of influential men he always sought to place
himself under some slight obligation; he bor-
rowed (and returned) a book, or asked some
small service. The obligation incurred was never
heavy enough to trouble *him*, but it always en-

couraged the other party to renewed bounty. The habit of generosity is apt to grow with exercise, and it is precisely because France was the friend and loyal ally of America upward of a century ago, that she is now ready, and always has been, to testify the warmth and fidelity of her attachment. And if there ever has been at any time, on the face of our friendship, coldness or estrangement, or the appearance of it, such a change has never been exhibited by France.

If I were called upon to pick out from the mass of concurring testimony proof of the priceless value of French aid to the American colonies, I should go to that dark and dreary winter at Valley Forge, when even the stoutest hearts were despondent. All that makes victory possible was absent, except courage and faith, and they were fast failing before the cruel blows of adverse fortune. What must other men have thought of the future and its promises when Washington from the midst of his shivering, half-clad, and half-fed followers, wrote this : " Unless some great and capital change takes place the army must be inevitably reduced to one or other of three things—starve, dissolve, or disperse."

Only a miracle could save the cause ! Who would help the struggling band of enthusiasts that had nothing to offer as a reward for the aid which they prayed for ? Was it not against all history and experience that the vanquished cause should so commend itself to the world that troops, and money, and friends, and sympathy from strangers—strangers in blood, in tastes, in language—should be provided as though a rich return were sure to follow? It all came, and strangely enough, the prime mover in the battle against monarchy was a king, the volunteers in the people's fight were nobles, the treasury that made success possible came from a well-nigh bankrupt state! If logic had had a voice in French councils, and French sentiment had not guided French action, Lafayette would have stayed at home, Louis XVI. would have closed his royal ear to these earnest appeals, French gold would have remained in French hands, and the galaxy of bright, brave, loyal, chivalrous marquises, dukes, and counts would never have fought, flirted, suffered, danced, and—died on American soil.

Frederic R. Coudert.

HARMOSAN.

Now the third and fatal conflict for the Persian
        throne was done,
And the Moslem's fiery valor had the crowning
        victory won.

Harmosan, the last and boldest the invader to
        defy,
Captive, overborne by numbers, they were bring-
        ing forth to die.

Then exclaimed that noble captive : " Lo, I perish in my thirst ;
Give me but one drink of water, and then arrive
 the worst !"

In his hand he took the goblet, but, awhile, the
 draught forbore,
Seeming doubtfully the purpose of the foeman
 to explore.

Well might then have paused the bravest—for,
 around him, angry foes
With a hedge of naked weapons did that lonely
 man enclose.

" But what fearest thou?" cried the caliph, " is
 it, friend, a secret blow ?
Fear it not !—our gallant Moslem no such treacherous dealing know.

" Thou may'st quench thy thirst securely, for
 thou shalt not die before
Thou hast drunk that cup of water—this reprieve
 is thine—no more !"

Quick the Satrap dashed the goblet down to
 earth with ready hand,
And the liquid sank for ever, lost amid the burning sand.

" Thou hast said that mine my life is, till the
 water of that cup
I have drained, then bid thy servants that spilled
 water gather up !"

For a moment stood the caliph as by doubtful
    passions stirred—
Then exclaimed, " For ever sacred must remain
    a monarch's word.

" Bring another cup, and straightway to the no-
    ble Persian give :
Drink, I said before, and perish—now I bid thee
    drink and live !"
                         Richard C. Trench.

THE RAVEN.

ONCE upon a midnight dreary, while I pondered,
    weak and weary,
Over many a quaint and curious volume of for-
    gotten lore—
While I nodded, nearly napping, suddenly there
    came a tapping,
As of some one gently rapping, rapping at my
    chamber-door.
" 'Tis some visitor," I muttered, " tapping at my
    chamber-door—
       Only this, and nothing more."

Ah, distinctly I remember, it was in the bleak
    December,
And each separate dying ember wrought its
    ghost upon the floor.
Eagerly I wished the morrow : vainly I had
    sought to borrow

From my books surcease of sorrow—sorrow for
    the lost Lenore—
For the rare and radiant maiden whom the an-
    gels name Lenore—
      Nameless here forevermore.

And the silken sad uncertain rustling of each
    purple curtain
Thrilled me, filled me with fantastic terrors
    never felt before;
So that now, to still the beating of my heart, I
    stood repeating,
" 'Tis some visitor entreating entrance at my
    chamber-door,—
Some late visitor entreating entrance at my
    chamber-door ;
      That it is, and nothing more."

Presently my soul grew stronger : hesitating then
    no longer,
" Sir," said I, " or Madam, truly your forgive-
    ness I implore ;
But the fact is I was napping, and so gently you
    came rapping,
And so faintly you came tapping, tapping at my
    chamber-door,
That I scarce was sure I heard you"—here I
    opened wide the door :
      Darkness there, and nothing more.

Deep into that darkness peering, long I stood
    there, wondering, fearing,

Doubting, dreaming dreams no mortals ever
    dared to dream before ;
But the silence was unbroken, and the stillness
    gave no token,
And the only word there spoken was the whis-
    pered word " Lenore ?"
This *I* whispered, and an echo murmured back
    the word, " Lenore!"
      Merely this, and nothing more.

Back into the chamber turning, all my soul
    within me burning,
Soon again I heard a tapping something louder
    than before.
"Surely," said I, " surely that is something at
    my window-lattice ;
Let me see, then, what thereat is, and this mys-
    tery explore,—
Let my heart be still a moment, and this mystery
    explore ;—
      'Tis the wind, and nothing more."

Open here I flung the shutter, when, with many
    a flirt and flutter,
In there stepped a stately raven of the saintly
    days of yore.
Not the least obeisance made he ; not a minute
    stopped or stayed he ;
But, with mien of lord or lady, perched above
    my chamber-door,—

Perched upon a bust of Pallas just above my
   chamber-door—
      Perched, and sat, and nothing more.

Then this ebony bird beguiling my sad fancy
   into smiling,
By the grave and stern decorum of the coun-
   tenance it wore,
"Though thy crest be shorn and shaven, thou,"
   I said, "art sure no craven,
Ghastly, grim, and ancient raven, wandering
   from the Nightly shore,
Tell me what thy lordly name is on the Night's
   Plutonian shore!"
      Quoth the raven, "Nevermore!"

Much I marvelled this ungainly fowl to hear discourse so plainly,
Though its answer little meaning—little rele-
   vancy bore;
For we cannot help agreeing that no living
   human being
Ever yet was blessed with seeing bird above his
   chamber-door—
Bird or beast upon the sculptured bust above
   his chamber-door,
      With such name as "Nevermore!"

But the raven sitting lonely on the placid bust,
   spoke only
That one word, as if his soul in that one word he
   did outpour.

Nothing further then he uttered—not a feather
  then he fluttered—
Till I scarcely more than muttered, " Other
  friends have flown before :
On the morrow *he* will leave me, as my hopes
  have flown before."
    Then the bird said, " Nevermore !"

Startled at the stillness broken by reply so
  aptly spoken,
" Doubtless," said I, " what it utters is its only
  stock and store,
Caught from some unhappy master, whom un-
  merciful disaster
Followed fast and followed faster, till his songs
  one burden bore,—
Till the dirges of his hope that melancholy burden bore,
    Of ' Never—nevermore !' "

But the raven still beguiling all my sad soul into
  smiling,
Straight I wheeled a cushioned seat in front of
  bird, and bust, and door ;
Then, upon the velvet sinking, I betook myself
  to linking
Fancy unto fancy, thinking what this ominous
  bird of yore—
What this grim, ungainly, ghastly, gaunt, and
  ominous bird of yore
    Meant in croaking " Nevermore !"

This I sat engaged in guessing, but no syllable
 expressing
To the fowl, whose fiery eyes now burned into
 my bosom's core ;
This and more I sat divining, with my head at
 ease reclining
On the cushion's velvet lining that the lamp-
 light gloated o'er,
But whose velvet violet lining with the lamp-
 light gloating o'er
 She shall press—ah, nevermore !

Then, methought, the air grew denser, perfumed
 from an unseen censer,
Swung by seraphim whose foot-falls tinkled on
 the tufted floor.
"Wretch," I cried, "thy God hath lent thee—by
 these angels he hath sent thee
Respite—respite and nepenthe from thy memories of Lenore !
Quaff, oh, quaff this kind nepenthe, and forget
 this lost Lenore !"
 Quoth the raven, "Nevermore !"

"Prophet !" said I, "thing of evil !—prophet
 still, if bird or devil !
Whether tempter sent, or whether tempest tossed
 thee here ashore,
Desolate, yet all undaunted, on this desert land
 enchanted—
On this home by Horror haunted—tell me truly,
 I implore—

Is there—*is* there balm in Gilead ?—tell me—tell
  me, I implore !"
      Quoth the raven, " Nevermore !"

" Prophet !" said I, " thing of evil !—prophet
  still, if bird or devil !
By that heaven that bends above us, by that
  God we both adore,
Tell this soul, with sorrow laden, if, within the
  distant Aidenn,
It shall clasp a sainted maiden whom the angels
  name Lenore ;
Clasp a rare and radiant maiden, whom the an-
  gels name Lenore !"
      Quoth the raven, "Nevermore !"

" Be that word our sign of parting, bird or
  fiend !" I shrieked, upstarting :
" Get thee back into the tempest and the Night's
  Plutonian shore !
Leave no black plume as a token of that lie thy
  soul hath spoken !
Leave my loneliness unbroken !—quit the bust
  above my door !
Take thy beak from out my heart, and take thy
  form from off my door !"
      Quoth the raven, " Nevermore !"

And the raven, never flitting, still is sitting, still
  is sitting
On the pallid bust of Pallas just above my
  chamber-door ;

And his eyes have all the seeming of a demon's
    that is dreaming,
And the lamp-light o'er him streaming throws
    his shadow on the floor;
And my soul from out that shadow that lies
    floating on the floor
        Shall be lifted—nevermore.
                    Edgar A. Poe.

MONEY MUSK.[1]

Ah, the buxom girls that helped the boys—
The nobler Helens of humbler Troys—
As they stripped the husks with rustling fold,
From eight-rowed corn as yellow as gold,

By the candle-light in pumpkin bowls,
And the gleams that showed fantastic holes
In the quaint old lantern's tatooed tin,
From the hermit glim set up within;

By the rarer light in girlish eyes
As dark as wells, or as blue as skies.
I hear the laugh when the ear is red,
I see the blush with the forfeit paid,

The cedar cakes with the ancient twist,
The cider cup that the girls have kissed.
And I see the fiddler through the dusk
As he twangs the ghost of "Money Musk!"

[1] Exercise on stress, median and thorough.

The boys and girls in a double row
Wait face to face till the magic bow
Shall whip the tune from the violin,
And the merry pulse of the feet begin.

MONEY MUSK.

In shirt of check, and tallowed hair,
The fiddler sits in the bulrush chair
Like Moses' basket stranded there
       On the brink of Father Nile.
He feels the fiddle's slender neck,
Picks out the notes with thrum and check,
And times the tune, with nod and beck,
       And thinks it a weary while.
All ready!   Now he gives the call,
Cries, " *Honor to the ladies !*"   All
The jolly tides of laughter fall
       And ebb in a happy smile.

D-o-w-n comes the bow on every string,
" *First couple join right hands and swing !*"
As light as any blue-bird's wing
       " *Swing once and a half times round.*"
Whirls Mary Martin all in blue—
Calico gown and stockings new,
And tinted eyes that tell you true,
       Dance all to the dancing sound.

She flits about big Moses Brown,
Who holds her hands to keep her down,
And thinks her hair a golden crown
       And his heart turns over once !

His cheek with Mary's breath is wet,
He gives a second somerset!
He means to win the maiden yet,
      Alas, for the awkward dunce!

"Your stoga boot has crushed my toe!"
"I'd rather dance with one-legged Joe!"
"You clumsy fellow!" "*Pass below!*"
      And the first pair dance apart.
Then "*Forward six!*" advance, retreat,
Like midges gay in sunbeam street.
'Tis Money Musk by merry feet
      And the Money Musk by heart!

"*Three quarters round your partner swing!*"
"*Across the set!*" The rafters ring,
The girls and boys have taken wing
      And have brought their roses out!
'Tis "*Forward six!*" with rustic grace,
Ah, rarer far than—"*Swing to place!*"
Than golden clouds of old point-lace,
      They bring the dance about.

Then clasping hands all—"*Right and left!*"
All swiftly weave the measure deft
Across the woof in loving weft,
      And the Money Musk is done!
Oh, dancers of the rustling husk,
Good-night, sweethearts, 'tis growing dusk,
Good-night for aye to Money Musk,
      For the heavy march begun!
               Benjamin F. Taylor.

THE BELL OF ATRI.

AT Atri in Abruzzo, a small town
Of ancient Roman date, but scant renown,
The Re Giovanni, now unknown to fame,
So many monarchs since have borne the name,
Had a great bell hung in the market-place
Beneath a roof, projecting some small space,
By way of shelter from the sun and rain.
Then rode he through the streets with all his
     train,
And, with the blast of trumpets loud and long,
Made proclamation, that whenever wrong
Was done to any man, he should but ring
The great bell in the square, and he, the King,
Would cause the Syndic to decide thereon.
Such was the proclamation of King John.

How swift the happy days in Atri sped,
What wrongs were righted need not here be said.
Suffice it that, as all things must decay,
The hempen rope at length was worn away,
Unravelled at the end, and, strand by strand,
Loosened and wasted in the ringer's hand,
Till one, who noted this in passing by,
Mended the rope with braids of briony,
So that the leaves and tendrils of the vine-
Hung like a votive garland at a shrine.

By chance it happened that in Atri dwelt
A knight, with spur on heel and sword in belt,
Who loved to hunt the wild-boar in the woods,
Who loved his falcons with their crimson hoods,

Who loved his hounds and horses, and all sports
And prodigalities of camps and courts;—
Loved, or had loved them; for at last, grown old,
His only passion was the love of gold.

He sold his horses, sold his hawks and hounds,
Rented his vineyards and his garden grounds,
Kept but one steed, his favorite steed of all,
To starve and shiver in a naked stall,
And day by day sat brooding in his chair,
Devising plans how best to hoard and spare.

At length he said: "What is the use or need
To keep at my own cost this lazy steed,
Eating his head off in my stables here,
When rents are low and provender is dear?
Let him go feed upon the public ways;
I want him only for the holidays."
So the old steed was turned into the heat
Of the long, lonely, silent, shadeless street;
And wandered in suburban lanes forlorn,
Barked at by dogs, and torn by brier and thorn.

One afternoon, as in that sultry clime
It is the custom in the summer time,
With bolted doors and window-shutters closed,
The inhabitants of Atri slept or dozed;
When suddenly upon their senses fell
The loud alarum of the accusing bell!
The Syndic started from his deep repose,
Turned on his couch, and listened, and then rose
And donned his robes, and with reluctant pace

Went panting forth into the market-place,
Where the great bell upon its cross-beam swung,
Reiterating with persistent tongue,
In half-articulate jargon, the old song:
"Some one hath done a wrong, hath done a
     wrong!"

But ere he reached the belfry's light arcade
He saw, or thought he saw, beneath its shade,
No shape of human form of woman born,
But a poor steed dejected and forlorn,
Who with uplifted head and eager eye
Was tugging at the vines of briony.
"Domeneddio!" cried the Syndic straight,
"This is the Knight of Atri's steed of state!
He calls for justice, being sore distressed,
And pleads his cause as loudly as the best."

Meanwhile from street and lane a noisy crowd
Had rolled together like a summer cloud,
And told the story of the wretched beast
In five-and-twenty different ways at least,
With much gesticulation and appeal
To heathen gods, in their excessive zeal.
The Knight was called and questioned; in reply
Did not confess the fact, did not deny;
Treated the matter as a pleasant jest,
And set at naught the Syndic and the rest,
Maintaining, in an angry undertone,
That he should do what pleased him with his
     own.

And thereupon the Syndic gravely read
The proclamation of the King; then said:
"Pride goeth forth on horseback grand and gay,
But cometh back on foot, and begs its way;
Fame is the fragrance of heroic deeds,
Of flowers of chivalry and not of weeds!
These are familiar proverbs; but I fear
They never yet have reached your knightly ear.
What fair renown, what honor, what repute
Can come to you from starving this poor brute?
He who serves well and speaks not, merits more
Than they who clamor loudest at the door.
Therefore the law decrees that as this steed
Served you in youth, henceforth you shall take
    heed
To comfort his old age, and to provide
Shelter in stall, and food, and field beside."

The knight withdrew abashed; the people all
Led home the steed in triumph to his stall.
The King heard and approved, and laughed in
    glee,
And cried aloud: "Right well it pleaseth me!
Church-bells at best but ring us to the door;
But go not in to mass; my bell doth more:
It cometh into court and pleads the cause
Of creatures dumb and unknown to the laws;
And this shall make, in every Christian clime,
The Bell of Atri famous for all time."

Longfellow.

THE LIFEBOAT.

You've heerd of the Royal Helen, the ship as was
  wrecked last year?
Yon be the rock she struck on—the boat as went
  out be here ;
The night as she struck was reckoned the worst
  as ever we had,
And this is a coast in winter where the weather
  be awful bad.
The beach here was strewed with wreckage, and
  to tell you the truth, sir, then
Was the only time as ever we'd a bother to get
  the men.
The single chaps was willin', and six on 'em vol-
  unteered,
But most on us here is married, and the wives
  that night was skeered.

Our women ain't chicken-hearted when it comes
  to savin' lives,
But death that night looked certain—and our
  wives be only wives ;
Their lot ain't bright at the best, sir ; but here,
  when the man lies dead,
'Tain't only a husband missin', it's the children's
  daily bread ;
So our women began to whimper and beg o' the
  chaps to stay—
I only heerd on it after, for that night I was
  kept away.
I was up at my cottage, yonder, where the wife
  lay nigh her end,

She'd been ailin' all the winter, and nothin' 'ud
    make her mend.

The doctor had given her up, sir, and I knelt by
    her side and prayed,
With my eyes as red as a babby's, that Death's
    hand might yet be stayed.
I heard the wild wind howlin', and I looked on
    the wasted form,
And thought of the awful shipwreck as had come
    in the ragin' storm ;
The wreck of my little homestead—the wreck of
    my dear old wife,
Who'd sailed with me forty years, sir, o'er the
    troublous waves of life,
And I looked at the eyes so sunken, as had been
    my harbor lights,
To tell of the sweet home haven in the wildest,
    darkest nights.

She knew she was sinkin' quickly, she knew as
    her end was nigh,
But she never spoke o' the troubles as I knew on
    her heart must lie,
For we'd had one great big sorrow with Jack, our
    only son—
He'd got into trouble in London, as lots o' the
    lads ha' done ;
Then he'd bolted, his masters told us—he was
    allus what folk call wild.
From the day as I told his mother, her dear face
    never smiled.

We heerd no more about him, we never knew
    where he went,
And his mother pined and sickened for the mes-
    sage he never sent.

I had my work to think of; but she had her
    grief to nurse,
So it eat away at her heartstrings, and her health
    grew worse and worse.
And the night as the Royal Helen went down on
    yonder sands,
I sat and watched her dyin', holdin' her wasted
    hands.
She moved in her doze a little, then her eyes
    were opened wide,
And she seemed to be seekin' somethin', as she
    looked from side to side;
Then half to herself she whispered, "Where's
    Jack, to say good-bye?
It's hard not to see my darlin', and kiss him
    afore I die!"

I was stoopin' to kiss and soothe her, while the
    tears ran down my cheek,
And my lips were shaped to whisper the words I
    couldn't speak,
When the door of the room burst open, and my
    mates were there outside
With the news that the boat was launchin'.
    "You're wanted!" their leader cried.
"You've never refused to go, John; you'll put
    these cowards right.

There's a dozen of lives maybe, John, as lie in our
    hands to-night!"
'Twas old Ben Brown, the captain ; he'd laughed
    at the women's doubt.
We'd allus been first on the beach, sir, when
    the boat was goin' out.

I didn't move, but I pointed to the white face on
    the bed—
"I can't go, mate," I murmured ; "in an hour
    she may be dead.
I cannot go and leave her to die in the night
    alone."
As I spoke Ben raised his lantern, and the light
    on my wife was thrown ;
And I saw her eyes fixed strangely with a plead-
    ing look on me,
While a tremblin' finger pointed through the
    door to the ragin' sea.
Then she beckoned me near, and whispered, "Go,
    and God's will be done !
For every lad on that ship, John, is some poor
    mother's son."

Her head was full of the boy, sir—she was think-
    ing, maybe, some day
For lack of a hand to help him his life might be
    cast away.
" Go, John, and the Lord watch o'er you ! and
    spare me to see the light,
And bring you safe," she whispered, " out of the
    storm to-night."

Then I turned and kissed her softly, and tried to
    hide my tears,
And my mates outside, when they saw me, set up
    three hearty cheers;
But I rubbed my eyes wi' my knuckles, and
    turned to old Ben and said,
"I'll see her again, maybe, lad, when the sea
    gives up its dead."

We launched the boat in the tempest, though
    death was the goal in view,
And never a one but doubted if the craft could
    live it through;
But our boat she stood it bravely, and, weary and
    wet and weak,
We drew in hail of the vessel we had dared so
    much to seek.
But just as we come upon her she gave a fearful
    roll,
And went down in the seethin' whirlpool with
    every livin' soul!
We rowed for the spot, and shouted,—for all
    around was dark,—
But only the wild wind answered the cries from
    our plungin' bark.

I was strainin' my eyes and watchin', when I
    thought I heerd a cry,
And I saw past our bows a somethin' on the crest
    of a wave dashed by;
I stretched out my hand to seize it. I dragged
    it aboard, and then

I stumbled, and struck my forrud, and fell like a
    log on Ben.
I remember a hum of voices, and then I knowed
    no more
Till I came to my senses, here, sir—here, in my
    home ashore.
My forrud was tightly bandaged, and I lay on my
    little bed—
I'd slipped, so they told me arter, and a rulluck
    had struck my head.

Then my mates came in and whispered ; they'd
    heerd I was comin' round.
At first I could scarcely hear 'em, it seemed like
    a buzzin' sound ;
But as soon as my head got clearer, and accus-
    tomed to hear 'em speak,
I knew as I'd lain like that, sir, for many a long,
    long week.
I guessed what the lads was hidin', for their poor
    old shipmate's sake.
I could see by their puzzled faces they'd got some
    news to break ;
So I lifts my head from the pillow, and I says to
    old Ben, " Look here !
I'm able to bear it now, lad—tell me, and never
    fear."

Not one on 'em ever answered, but presently Ben
    goes out,
And the others slinks away like, and I says,
    " What's this about ?

Why can't they tell me plainly as the poor old
    wife is dead?"
Then I fell again on the pillows, and I hid my
    achin' head.
I lay like that for a minute, till I heard a voice
    cry " John !"
And I thought it must be a vision as my weak
    eyes gazed upon ;
For there by the bedside, standin' up and well,
    was my wife.
And who do ye think was with her ? Why,
    Jack, as large as life.

It was him as I'd saved from drownin' the night
    as the lifeboat went
To the wreck of the Royal Helen ; 'twas that as
    the vision meant.
They'd brought us ashore together, he'd knelt by
    his mother's bed,
And the sudden joy had raised her like a miracle
    from the dead ;
And mother and son together had nursed me
    back to life,
And my old eyes woke from darkness to look on
    my son and wife.
Jack ? He's our right hand now, sir ; 'twas Provi-
    dence pulled him through—
He's allus the first aboard her when the lifeboat
    wants a crew.

                      George R. Sims.

COURTSHIP OF HENRY THE FIFTH.

King Henry.   Fair Katherine and most fair !
Will you vouchsafe to teach a soldier terms,
Such as will enter at a lady's ear,
And plead his love-suit to her gentle heart?

Katherine (conversing with the aid of Alice.) Your
majesty shall mock at me ; I cannot speak your
England.

K. Hen.   O fair Katherine, if you will love me
soundly with your French heart, I will be glad to
hear you confess it brokenly with your English
tongue.   Do you like me, Kate ?

Kath.   *Pardonnez moi,* I cannot tell vat is—like
me.

K. Hen.   An angel is like you, Kate ; and you
are like an angel.

Kath.   *Que dit-il ? que je suis semblable à les
anges ?*

Alice.   *Oui, vraiment (sauf votre grace), ainsi
dit-il.*

K. Hen.   I said so, dear Katherine, and I must
not blush to affirm it.

Kath.   *O bon Dieu ! les langues des hommes sont
pleines de tromperies.*

K. Hen.   What says she, fair one? that the
tongues of men are full of deceits?

Alice.   *Oui ;* dat de tongues of de mans is full
of deceits : dat is de princess.

K. Hen.   The princess is the better English-
woman.   I' faith, Kate, my wooing is fit for thy
understanding.

Kath. *Sauf votre honneur*, me understand well.

K. Hen. Marry, if you would put me to verses, or to dance for your sake, Kate, why you undid me. If I could win a lady at leap-frog, or by vaulting into my saddle with my armor on my back, under the correction of bragging be it spoken, I should quickly leap unto a wife. Or if I might buffet for my love, or bound my horse for her favors, I could lay on like a butcher, and sit like a jackanapes, never off; but, Kate, I cannot look greenly, nor gasp out my eloquence, nor I have no cunning in protestation. If thou canst love a fellow of this temper, Kate, whose face is not worth sunburning, that never looks in his glass for love of anything he sees there, let thine eye be thy cook. I speak to thee plain soldier: If thou canst love me for this, take me; if not, to say to thee—that I shall die, is true: but—for thy love, by the Lord, no; yet I love thee too. And while thou livest, dear Kate, take a fellow of plain and un-coined constancy; for he perforce must do thee right, because he hath not the gift to woo in other places: for these fellows of infinite tongue, that can rhyme themselves into ladies' favors, they do always reason themselves out again. What! a speaker is but a prater; a rhyme is but a ballad. A good leg will fall; a straight back will stoop; a black beard will turn white; a curled pate will grow bald; a fair face will wither; a full eye will wax hollow; but a good heart, Kate, is the sun and the moon; or rather the sun, and not the moon; for it shines bright, and never changes, but keeps his course truly. If thou wouldst have such a one, take me.

Kath. Is it possible dat I should love de enemy of France?

K. Hen. No; it is not possible you should love the enemy of France, Kate; but in loving me, you should love the friend of France; for I love France so well that I will not part with a village of it; I will have it all mine; and, Kate, when France is mine, and I am yours, then yours is France, and you mine.

Kath. I cannot tell vat is dat.

K. Hen. No, Kate? I will tell thee in French; which, I am sure, will hang upon my tongue like a new-married wife about her husband's neck, hardly to be shook off. *Quand j'ai la possession de France, et quand vous avez la possession de moi* (let me see, what then? Saint Dennis be my speed!)—*donc votre est France, et vous êtes mien.* It is as easy for me, Kate, to conquer the kingdom as to speak so much more French: I shall never move thee in French, unless it be to laugh at me.

Kath. *Sauf votre honneur, le Français que vous parlez est meilleur que l'Anglais le quel je parle.*

K. Hen. No, 'faith, is't not, Kate; but thy speaking of my tongue, and I thine, must needs be granted to be much at one. But, Kate, dost thou understand thus much English? Canst thou love me?

Kath. I cannot tell.

K. Hen. Can any of your neighbors tell, Kate?

I'll ask them. Come, I know thou lovest me.
How answer you, *la plus belle Katherine du monde,
mon tres chere et divine déesse?*

Kath. Your *majesté* ave *fausse* French enough
to deceive de most *sage demoiselle* dat is *en France.*

K. Hen. Now, fie upon my false French! By
mine honor, in true English, I love thee, Kate.
Put off your maiden blushes; avouch the thoughts
of your heart with the looks of an empress; take
me by the hand, and say—Harry of England, I
am thine: which word thou shalt no sooner bless
mine ear withal but I will tell thee aloud—England is thine, Ireland is thine, France is thine,
and Henry Plantagenet is thine; who, though I
speak it before his face, if he be not fellow with
the best king, thou shalt find the best king of
good fellows. Come, your answer in broken
music; for thy voice is music, and thy English
broken. Wilt thou have me?

Kath. Dat is as it shall please de *roi mon pere.*

K. Hen. Nay, it will please him well, Kate;
it shall please him, Kate.

Kath. Den it sall also content me.

K. Hen. Upon that I kiss your hand, and I
call you my queen.

*Kath. Laissez, mon seigneur, laissez, laissez; ma
foi, je ne veux point que vous abbaissez votre grandeur, en baisant la main d'une votre indigne serviteure;
excuzez moi, je vous supplie, mon tres puissant
seigneur.*

K. Hen. Then I will kiss your lips, Kate.

Kath. Les dames, et demoiselles, pour etre bais-

sées devant leur noces, il n'est pas le coutûme de France.

K. Hen. Madam my interpreter, what says she ?

Alice. Dat it is not be de fashion *pour les* ladies of France—I cannot tell what is *baiser, en* English.

K. Hen. To kiss.

Alice. Your majesty *entendre* bettre *que moi.*

K. Hen. It is not the fashion for the maids in France to kiss before they are married, would she say ?

Alice. Oui, vraiment.

K. Hen. O Kate, nice customs curt'sy to great kings. Dear Kate, you and I cannot be confined within the weak list of a country's fashion ; therefore, patiently and yielding. [*Kissing her.*] You have witchcraft in your lips, Kate ; there is more eloquence in a sugar touch of them than in the tongues of the French council : and they should sooner persuade Henry of England than a general petition of monarchs.

Shakespeare.

DAISIES.

WERE I but a nymph or fairy,
Full of whims and fancy-free,
Of all form in which to hide me,
I would fain a daisy be :
Ever fresh on lawn or meadow,
Ready with its smile to greet   ·
Rain or sunshine, bird or dewdrop,
Or the tread of human feet.

All through life to us poor mortals
  Faithful friends the daisies are ;
Children on the green grass playing
  Hail their bright eyes as a star.
Decked with chains of daisy-blossoms,
  Up and down they go like kings,
Call its golden eyes their money,
  Cast abroad its silver wings.

Older grown, upon the green turf
  Many a measure light they tread,
Ah ! so light the friendly daisy
  Lifts unhurt its little head :
Days of music and of beauty,
  With a golden halo bright ;
How could hearts so gay and happy
  Fail to tread with footstep light ?

Heavier soon that footstep presses,
  Oh, the weary, weary day !
What to them the smile of April ?
  What to them the bloom of May ?
'Neath the hawthorn in the meadow
  They would fling them down and die ;
In the turf that cools their forehead
  Greets them still the daisy's eye.

Like a friend it cheers and soothes them,
  Brings their childhood back again,
Till upon the daisy-blossoms
  Fall their tears like welcome rain.
Lowly lessons has it whispered,
  Smoothing still the sullen frown ;

Not for sorrow nor for anguish,
  Must we lay our burden down.

Life rolls on, and ever onward,
  All things round them fleet and change;
Home and childhood both have vanished:
  They are old, and all is strange;
Only in the early spring-time
  One familiar sight they see,
On the lawn the starry daisy
  Bright as it was wont to be.

Life rolls on, and ever onward,
  Friendly death has come at last;
Weep not, for indeed 'twas welcome,
  And with sign of faith they passed.
With the Church's prayers and blessings
  Laid to rest in holy ground,
All of earth their mem'ry keeping
  Is one little daisied mound.

O ye daisies! well ye teach us
  Friendship's holy debt to pay,
'Neath the mower's footstep springing,
  Growing by the common way:
Moralists would fain persuade us
  Kindly hearts are few and rare;
Never can I learn the lesson,
  For I find them everywhere.

Up and down our pathway scattered,
  Like the daisies, do they lie;
We need only glance around us

Would we meet their friendly eye.
Pride indeed may overlook them,
   Cold contempt may turn aside,
But, believe me, daisies ever
   In the grass their blossoms hide.

Childlike hands will seek and find them,
   Childlike hearts the treasure prize ;
Never will they scorn the welcome
   Which they read in daisy's eyes.
Smile not ; could I choose my calling,
   Nothing sweeter could I see
Than to all sad hearts and lonely
   Might I like the daisy be !
               "Songs in the Night."

THE DEATH OF D'ASSAS.

[In the autumn of 1760, Louis XV. sent an army into Germany. They took up a strong position at Klostercamp, intending to advance on Rheinberg. The young Chevalier D'Assas was sent out by Auvergne to reconnoitre. He met a party advancing to surprise the French camp. Their bayonets pricked his breast, and the leader whispered, "Make the least noise, and you are a dead man." D'Assas paused a moment, then cried out as loud as he could, "Here, Auvergne! here are the enemy!" He was immediately cut down, but his death had saved the French army.—*History of France.*]

THERE'S revelry at Louis' court. With joust and
   tournament,
With feasting and with laughter, the merry days
   are spent ;
And midst them all, those gallant knights, of
   Louis' court the boast,

Who can compare with D'Assas among the brill-
    iant host?
The flush of youth is on his cheek; the fire that
    lights his eye
Tells of the noble heart within, the spirit pure
    and high.
No braver knight holds charger's reign, or
    wields the glittering lance,
Than proud and lordly D'Assas, bold chevalier of
    France.

The sound of war strikes on the air from far be-
    yond the Rhine,
Its clarions ring across the fields, rich with the
    purple vine.
France calls her best and bravest: " Up, men,
    and take the sword !
Of German vales and hillsides, Louis would fain
    be lord;
Go forth, and for your sovereign win honor and
    renown;
Plant the white flag of Ivry on valley and on
    town.
The green soil of the Fatherland shall see your
    arms advance,
The dull and stolid Teuton shall bend the knee
    to France."

On Klostercamp the morning sun is glancing
    brightly down.
Auvergne has ranged his forces within the an-
    cient town.

From thence on Rheinberg shall they move:
    that citadel so grim
Shall yield her towers to Auvergne, shall ope her
    gates to him.
His warriors stand about him, a bold and gallant
    band,
No general e'er had truer men to follow his com-
    mand.
He seeks the best and bravest; on D'Assas falls
    his glance,—
On brave and lordly D'Assas, bold chevalier of
    France.

" Advance, my lord," cried Auvergne; D'Assas is
    at his side.
" Of all the knights who form my train, who
    'neath my banner ride,
None hold the place of trust the king our sov-
    ereign gives to thee,—
Wilt thou accept a fearful charge that death or
    fame shall be?
Wilt thou, O D'Assas! ride to-night close to
    the foeman's line,
And see what strength he may oppose to these
    proud hosts of mine ? "
Then D'Assas bows his stately head. " Thy will
    shall soon be done.
Back will I come with tidings full e'er dawns the
    morning sun."

'Tis midnight. D'Assas rideth forth upon his
    well-tried steed.

Auvergne hath made a worthy choice for this adventurous deed.

But stop! what means this silent host? How stealthily they come!

No martial music cleaves the air, no sound of beaten drum.

Like spectre forms they seem to glide before his wondering eyes;

Well hath he done, the wary foe, to plan this wild surprise.

Back D'Assas turns; but ah! too late,—a lance is laid in rest:

The knight can feel its glittering point against his corselet prest.

"A Frenchman! Hist!" A heavy hand has seized his bridle-rein.

"Hold close thy lips, my gallant spy; one word, and thou art slain.

What brought thee here? Dost thou not know this is the Fatherland?

How dar'st thou stain our righteous earth with thy foul Popish band?

Wouldst guard thy life, then utter not one sound above thy breath;

A whisper, and thy dainty limbs shall make a meal for Death.

Within thy heart these blades shall find the black blood of thy race,

And none shall ever know or dream of thy last resting-place."

Calm as a statue D'Assas stands. His heart he lifts on high.

"The God of battles help me now, and teach
    me how to die!
A weeping maid will mourn my fate, a sovereign
    holds me dear ;
Be to them ever more than I who perish sadly
    here."
No word has passed his pallid lips, no sound his
    voice has made.
'Twas but the utterance of his heart, this prayer
    the soldier prayed.
But then? ah, then! No voice on earth e'er
    rang more loud and clear:
"Auvergne!" he cried, "Auvergne, Auvergne!
    Behold! the foe is here!"

The forest echoes with the shout. Appalled his
    captors stand.
The courage of that dauntless heart has stayed
    each murderous hand.
A moment's pause,—then who can tell how quick
    their bayonets' thrust
Reached D'Assas' heart, and laid him there, a
    helpless heap of dust!
The bravest chevalier of France, the pride of
    Louis' train,—
His blood bedews that alien earth, a flood of
    crimson rain.
But Auvergne—Auvergne hears the cry; his
    troops come dashing on :
Ere D'Assas' spirit leaves its clay, the victory
    has been won.

          Mary E. Vandyne, in " Good Cheer."

JACQUES DUFOUR.

STROLLING in the cool of evening, drinking in the
    balmy air,
·I met a strange wayfaring man bowed down with
    grief and care.
Eighty years had left their foot-prints on his
    gaunt and ashen cheek,
And his hands were gray and shrunken, and his
    voice was thin and weak :
But his eyes, while he was speaking, kindled with
    a misty glow,
'Mid their whitened brows and lashes, like a cra-
    ter in the snow.
And this aged Frenchman told me (his name was
    Jacques Dufour),
The story of the faded shred of ribbon that he
    wore :
Just a scrap of scarlet ribbon pinned upon his
    shrunken breast,
But to him more rich and beautiful than rubies
    of the East.
'Twas in eighteen-twelve he won it, in that ter-
    rible campaign
When the French invaded Russia, but invaded
    her in vain ;
And the starved and freezing Frenchmen had
    begun that sad retreat
Through the snow that proved for most of them
    both grave and winding-sheet.
There had been a bloody skirmish 'twixt the rear-
    guard and the foe,

And among the sorely wounded, whom the chance
    of fight laid low,
Was a gallant Polish Colonel, Marshal Davoust's
    favorite aide,
And the Marshal, kneeling o'er him, turned about,
    and sharply said :
"Halt, Company of Grenadiers, and see this
    wounded Pole !
He loves the French; he hates the Russ, with
    all his fiery soul :
Will you let him fall a prisoner to his bloody-
    minded foe?"
And the Company of Grenadiers cried out as one
    man, *No!*
"Then lift him," said the Marshal. "You sol-
    diers must have learned
That our wagons we've abandoned, and our bag-
    gage has been burned :
Make a litter; you must bear him; I trust him
    to your love ;
He will burden, will impede you, but I know that
    you will prove
That you do your duty ever, and will guard this
    wounded man
As you guard your sacred colors when they lead
    the battle's van."
So they made the hasty litter, and the wounded
    man they bore
(Of the youngest and the cheeriest, was Sergeant
    Jacques Dufour).
And day by day they fought their way, through
    deserts bleak and wild,

Guarding the crippled Colonel, as a woman
    guards her child.
But the work of love delayed them, and they
    slowly fell behind,
Yet not one of all that Company of Grenadiers
    repined.
Still they fought the cold and Cossacks; still
    they held their rugged way,
Falling back, but never fleeing: retreating, yet
    at bay.
But the foe was fell and agile, and the cold it
    waxed amain,
And so one by one they perished—some were
    frozen, some were slain,
Till the nineteenth day of marching came, and
    there were only five
Of that Company of Grenadiers who still re-
    mained alive.
Then spoke the wounded Colonel: "Oh, my
    comrades, it is vain:
I can surely never live to see my native land again;
You are squandering your lives for nought, lives
    it were sweet to save
For France and future glory; so leave me, com-
    rades brave."
"*Peste!*" said Jacques Dufour, "my Colonel, we
    take leave to answer *Nay*.
We have orders to deliver you at Wilna,—we
    obey!"
So they lift again the litter, and they struggle on
    their way,

Till the western clouds are lighted with the
    gleams of dying day:
And as they watch the glory, against those golden
    skies,
The towers and walls of Wilna in welcome out-
    line rise!
But too great the stress of feeling for those over-
    burdened men:
Too swift the refluent flood of hope that swelled
    their hearts again:
Far too weak their feeble bodies for this beatific
    sight:
Two fell dying on the left hand, two fell dying
    on the right;
And, as faded in the frozen air their last convul-
    sive moan,
Lo! of all that noble Company, Dufour was left
    alone!
Did he falter? No! He lifted in his arms the
    wounded man,
And with wild and desperate shouting towards
    the nearest outpost ran;
And the pickets came with succor, and the sun
    had just gone down
When they bore the Sergeant and his charge in
    safety to the town.
Then Dufour sent up a message to headquarters,
    quaint and short,
That " the Company of Grenadiers desired to re-
    port."
"Granted," said the bluff old Marshal, " let
    them do it here and now."

And Jacques Dufour came marching in and made
    his stiffest bow.
" Where is my wounded Colonel?" " Safe in
    the hospital,
Where you ordered us to place him, Monsieur
    le Marechal."
" Where is the Company? They too have come
    in safety all?"
" The Company is present, Monsieur le Mare-
    chal."
" Where is the Company, I repeat, the *Company* ?"
    once more.
" The Company is *present*," said Sergeant Jacques
    Dufour.
" But your comrades—there were ninety or a
    hundred men, you know."
" Ah, mon Marechal, my comrades lie buried in
    the snow!"
Then up rose the stout old Marshal, with his
    eyes brimful of tears,
Dashed aside the barriers of rank, the cold re-
    serve of years :
Caught the stripling to his bosom, gave him a
    reverent kiss,
And the ribbon which Dufour has worn from
    that far day to this.
William W. Howe.

THE SIRENS.

SWEETLY they sang in the days of old,
    Till the mariners heard them far at sea,
And, lured by that music, the brave and bold,
    Buffeting billows wild and free,

Forgot their duty, and shifting sail,
　　Steered to the treacherous music's fall;
Ah! better have battled the sharpest gale,
　　Than lent the ear to the Sirens' call.

For, bleaching bare on the cold, white sand,
　　Lo! countless victims, who bent an oar
From the safe, strong waves, to the false, fair
　　　land,
　　And perished there on the cruel shore.

You say no longer the Sirens sing,
　　And cheat the souls of the sons of men;
That over life's breakers no harp-notes ring
　　With perilous sweetness fraught, as when
In the gray, dim dawn of the waking world
　　The sailors leaned from the decks to hear
Those wooing strains, till their flag they furled,
　　And sped to the tempters who cost them dear.

Be not too sure! Till the lips are dumb,
　　And the brow is chill in the damp of death,
There are always Sirens to overcome,
　　And their tones are sweet as a bugle's breath.

Who faints and falters, in heart and hand,
　　When nights are dreary and storms are cold,
Who hears, as if by the zephyrs fanned,
　　False love-notes blown, as in days of old,
Who barters his hope of the peace of God
　　For a present ease, a delusive rest,
Is treading the path that is always trod
　　By feet astray from the steadfast best.

And the mocking Sirens, who comb their locks
    And weave their charms for the foolish heart
Till it breaks itself on the sunken rocks,
    Still smile and sing with a fatal art.

Who spends his money before 'tis earned;
    Who covets the splendor he cannot buy;
Who silently listens when good is spurned;
    For the coin of honor, who gives a lie;
Who, weak of armor, does not endure,
    When the conflict deepens, and wounds are
        felt;
The man, whose soul is no longer pure
    As when at his mother's knee he knelt,

Has heard where the white-caps kiss the reef,
    The baleful strain that the Sirens sing;
Though his joy be bright, it shall still be brief,
    And the hateful harps shall his death-knell
        ring.

You may stop your ears as you sail along,
    And drift away from their misty coast;
Or better still, you may lift a song
    That is sweeter than theirs, for all their boast.
That song shall soar to the heights above,
    And thence, like a silver star, shall fall
To hearken and cheer, with tones of love,
    All souls that list to its dulcet call.

In vain do the Sirens sing for one
    Whose spirit is tuned to higher praise,
And who meekly fills, with duties done,
    The rounded spheres of life's common days.
        Margaret E. Sangster, in " Good Cheer."

THE CONVICT SHIP.

HE stood upon the emerald strand,
  Which gave his sires a kingly birth,
An exile from his native land,
  A sentenced wanderer o'er the earth.
No, not the earth! or he had found
  A spot where free-born souls could dwell.

Like common felons, he was bound
  A slave beyond the ocean's swell—
The slave of England's haughty power—
  A branded convict by the laws—
The hero of one glorious hour—
  The martyr of a deathless cause.

Away upon the shadowy stream
  A prison ship, with spectral shrouds,
Like phantom vessel, in a dream,
  Floated between the waves and clouds;
And friends looked sternly on the deep,
  For kindred blood rose wild and high.

And grief so proud it might not weep,
  Burned fiercely in each lurid eye;
Thus, stung with grief and mute with rage,
  That band drew close to where he stood,
The victim of a faithless age,
  Sublime in their firm brotherhood.

"Advance!" The fatal word is given;
  A sob goes swelling through the crowd;
He lifts his trembling hands to heaven;
  His voice is mournful, deep, and loud.

"O God! I sought but to be free!
  If the deep bondage of this land
May centre all its ills on me,
  Then let me perish where I stand!
The blood of many a kingly sire
  Has reddened on my native sod;
The light of many a martyr's fire
  Has sent its life-smoke up to God!

"And I, her son, was it a crime
  To seize the chains that mar her breast,
And scatter back the wrongs that Time
  Has rusted round her emerald crest?

"A crime! While Ireland in her chains
  Against oppression toils and strives,
Each ruddy drop within my veins,—
  Though vital with a thousand lives,—
Let forth by this too willing hand
  If that could rend one link apart,
Should redden down this thirsty sand,
  The old wine of a broken heart.
Ireland, my country, what is she?
  And what am I? A convict slave!
An hour, and that remorseless sea
  Will bear me to a felon's grave."
"Onward!" The guards file slowly past;
  His pulse beats like a muffled knell;
As dead leaves in the wintry blast,
  His lifted hands unlocked, and fell.

But hark! A tumult in the crowd—
  Murmurs of anguish and surprise;

For onward, like a drifting cloud,
   That from a tempest wildly flies,—
The wife appears! A little child
   Lies struggling in her firm embrace,
And lifts his eyes so wide and wild,
   In terror, to her pallid face.
She knows not that her brow is bare,
   Nor feels the moist wind wander through
The golden splendor of her hair,
   That shades those eyes of burning blue—
Nor heeds the boy, but firmer girds
   His cries and struggles to her heart.
She utters neither groan nor words,
   But, with white cheek and lips apart,
Moves slowly through the breathless throng,
   That yields, with sympathy profound,
A passage, as she glides along
   In search of that brave outward bound.

The convict sees her on the strand;
   With one great pang of more than joy,
He turns upon the soldiers' band:
   "Stop! Yonder are my wife and boy!"
Then, like a panther from its glade,
   He braves the bayonets' deadly clash,
And flings aside each gleaming blade
   With a fierce bound and lightning dash.
She sees him. Like a wounded doe,
   All wild with bliss and mad with pain,

Springs to his arms: "I go—I go!
   What power shall part us two again?

Yes, fold us closer—closer, love—
  In me thou hast old Ireland yet;
I tell thee any land shall prove
  Native to us. Thy eyes are wet—
This great heart swells against my own,
  Its holy anguish pleads for me—
Ah! could ye leave us here alone,
  To perish looking on that sea?"

Breathless she gazed into his face,
  And lifted from her heart the child—
Relieved from her too fond embrace,
  The boy looked up and softly smiled.
The convict turned. *They* should not know
  How close the tears rose to his eyes—
How sweet the love—how deep the woe,
  Brought to his soul by this surprise.

"Advance!" Again that dread command
  Rolls, trumpet-like, along the shore;
Guarded by England's soldier band,
  They leave old Ireland evermore.

They stood together on that gloomy deck,
  Straining each gaze to catch another sight
Of that dim shore that, like a cloudy speck,
  Lay dark and sombre in the morning light.
Ah! it was very mournful. All around
  The weltering sea heaved with a hollow moan,
And from the hold arose the dreary sound
  Of smothered tears and many a broken groan;

For that old battered prison-ship was full,
  And freighted deep with misery and tears,
From the tall spars down to the creaking hull
  She reeled and trembled as with human fears.
They stood together, silently and still,
    Their eyes turned shoreward, with a dreary
      gaze ;
The winds swept wailing by them, fierce and
    chill,
  And there lay Ireland, mourning in the haze.
Then all at once his mustering grief awoke :
  " Oh, for a grave beneath my native sod !"
Thus on the wailing air his anguish broke,
  " I ask but this—but this, Almighty God !"

As if the heavens themselves had heard
  The passion of that cry,
The moaning deep was fiercely stirred ;
  The waves rose white and high.

The clouds, in one broad thunder-fold,
  Swept blackly through the air,
Like a great pall of death unrolled
  By angels answering there.

Blacker and blacker glooms the front of heaven,
  And in its wrath sweeps the recoiling sea ;
Hither and yon the angry waves are driven,
  Like routed war steeds, rushing to be free.
In mad battalions, with their white manes stream-
      ing,

They trample down the bosom of the deep ;
Sharp, fiery lances through the clouds are gleam-
    ing,
  And strike the waters, where they foam and
    leap.

And then was torn that inky cloud,
  With storms of lurid rain ;
And heaven's artillery thundered loud
  Above the heaving main.
It seemed as if the stars at last,
  Retreating in their ire,
Had poured upon the raging blast
  Great cataracts of fire.

Like a wild desert steed, beneath the lash,
  The tortured vessel plunges madly on ;
The masts have fallen with a smothered crash,
  Her guards are broken and her strength is
    gone.
The waves leap, rioting, across her deck,
  The helpless crew are swept from where they
    clung,
With a faint death-hold, to the plunging wreck,
  Her tattered canvas to the storm is flung.

Onward, still onward, anchorless and bare,
  She reels and toils toward the rocky shore ;
Her hold sends forth its shrieks of fell despair,
  Like fiery arrows piercing through the roar.

The exiles kneel together.—His embrace
  Girds in the unconscious boy and pallid wife,
A mighty gladness brightens on his face,
  Hope comes with death and slavery with life.

It comes! it comes! that rushing mountain, now,
  And lifts the shuddering vessel on its breast.
Quick, vivid flashes curl around her prow,
  And wreathe the masts with many a fiery crest.
A plunge, a quick recoil, one fearful cry!
  She strikes—she strikes—the rock—O God—
    the rock!
Amid the waters raging to the sky,
  That clinging group go downward with the
    shock.

'Tis over—from behind that parted cloud
  The frightened moon sheds down a timid
    gleam.
With the white foam around her, like a shroud,
  Through which her golden locks all dimly
    stream,
That gentle mother clasps her lifeless child,
  Folded upon the marble of her breast.
There in his pallid death the infant smiled,
  As if he lay caressing and caressed.

With his cold face half veiled beneath her hair,
  Cast to her side by a relenting wave,
The sire and husband had his answered prayer—
  The Irish Patriot filled an Irish Grave.

            Mrs. Ann S. Stephens.

THE LADY OF CASTLENORE.

(A.D. 1700.)

BRÉTAGNE had not her peer.   In the Province far
   or near,
   There were never such brown tresses, such a
      faultless hand :
She had youth, and she had gold, she had jewels
   all untold,
   And many a lover bold wooed the Lady of the
      Land.

But she with queenliest grace, bent low her pal-
   lid face,
   And, " Woo me not for mercy's sake, fair gentle-
      men," she said.
If they wooed then, with a frown she would
      strike their passion down :
   She might have wed a crown to the ringlets on
      her head.

From the dizzy castle-tips, hour by hour she
      watched the ships,
   Like sheeted phantoms coming and going ever-
      more,
While the twilight settled down on the sleepy
      seaport-town,
   On the gables peaked and brown, that had
      sheltered kings of yore.

Dusky belts of cedar-wood partly clasped the
   widening flood ;

Like a knot of daisies lay the hamlets on the
hill ;
In the hostelry below, sparks of light would
come and go,
And faint voices strangely low, from the garru-
lous old mill.

Here the land in grassy swells gently broke;
there sunk in dells
With mosses green and purple, and prongs of
rock and peak ;
Here in statue-like repose, an old wrinkled mountain rose,
With its hoary head in snows, and wild roses
at its feet.

And so oft she sat alone in the turret of gray
stone,
And looked across the moorland, so woful, to
the sea,
That there grew a village-cry, how her cheek did
lose its dye,
As a ship, once, sailing by, faded on the sap-
phire lea.

Her few walks led all one way, and all ended at
the gray,
And ragged, jagged rocks that fringe the lonesome beach ;
There she would stand, the Sweet ! with the
white surf at her feet,
While above her wheeled the fleet sparrow-
hawk with startled screech.

And she ever loved the sea, God's half-uttered
  mystery,
  With its million lips of shells, its never-ceas-
    ing roar;
And 'twas well that, when she died, they made
  her a grave beside
  The blue pulses of the tide, by the towers of
    Castlenore.

Now, one chill November morn, many russet
  autumns gone,
  A strange ship with folded wings lay dozing
    off the lea;
It had lain throughout the night with its wings
  of murky white
  Folded, after weary flight, the worn nursling of
    the sea.

Crowds of peasants flocked the sands, there were
  tears and clasping hands;
  And a sailor from the ship stalked through the
    kirk-yard gate;
Then amid the grass that crept, fading over her
  who slept,
  How he hid his face and wept, crying, *Late,
    alas! too late!*

And they called her cold. God knows. Underneath the winter snows,
  The invisible hearts of flowers grow ripe for
    blossoming!

And the lives that look so cold, if their stories
　　could be told,
Would seem cast in gentler mould, would
　　seem full of love and spring.
　　　　　　　　　　　　　T. B. Aldrich.

THE KING AND THE CHILD.

THE sunlight shone on the walls of stone
　　And towers sublime and tall ;
King Alfred sat upon his throne
　　Within his council hall.

And glancing o'er the splendid throng,
　　With grave and solemn face,
To where his noble vassals stood,
　　He saw a vacant place.

" Where is the Earl of Holderness ?"
　　With anxious look, he said.
" Alas, O King !" a courtier cried,
　　" The noble Earl is dead !"

Before the monarch could express
　　The sorrow that he felt,
A soldier with a war-worn face
　　Approached the throne and knelt.

"My sword," he said, " has ever been,
　　O King ! at thy command,
And many a proud and haughty Dane
　　Has fallen by my hand.

"I've fought beside thee in the field,
  And 'neath the greenwood tree;
It is but fair for thee to give
  Yon vacant place to me."

"It is not just," a statesman cried,
  "This soldier's prayer to hear,
My wisdom has done more for thee
  Than either sword or spear.

"The victories of the council hall
  Have made thee more renown
Than all the triumphs of the field
  Have given to thy crown.

"My name is known in every land,
  My talents have been thine,
Bestow this Earldom, then, on me,
  For it is justly mine."

Yet, while before the monarch's throne
  These men contending stood,
A woman crossed the floor who wore
  The weeds of widowhood.

And slowly to King Alfred's feet
  A fair-haired boy she led—
"O King! this is the rightful heir
  Of Holderness," she said.

"Helpless he comes to claim his own,
  Let no man do him wrong,
For he is weak and fatherless,
  And thou art just and strong."

"What strength of power," the statesman cried,
  "Could such a judgment bring?
Can such a feeble child as this
  Do aught for thee, O King,

"When thou hast need of brawny arms
  To draw thy deadly bows,
When thou art wanting crafty men
  To crush thy mortal foes?"

With earnest voice the fair young boy
  Replied: "I cannot fight,
But I can pray to God, O King!
  And Heaven can give thee might!"

The King bent down and kissed the child,
  The courtiers turned away.
"The heritage is thine," he said,
  "Let none their right gainsay.

"Our swords may cleave the casques of men,
  Our blood may stain the sod,
But what are human strength and power
  Without the help of God!"

Eugene J. Hall.

THE CHILDREN'S CRUSADE.

I.

FIRST VOICE.

WHAT is this I read in history,
Full of marvel, full of mystery,
Difficult to understand?
·Is it fiction, is it truth?

Children in the flower of youth,
Heart in heart, and hand in hand,
Ignorant of what helps or harms,
Without armor, without arms,
Journeying to the Holy Land!

Who shall answer or divine?
Never since the world was made
Such a wonderful crusade
Started forth for Palestine.
Never while the world shall last
Will it reproduce the past;
Never will it see again
Such an army, such a band,
Over mountain, over main,
Journeying to the Holy Land.

Like a shower of blossoms blown
From the parent trees were they;
Like a flock of birds that fly
Through the unfrequented sky,
Holding nothing as their own,
Passed they into lands unknown,
Passed to suffer and to die.

O the simple child-like trust!
O the faith that could believe
What the harnessed, iron-mailed
Knights of Christendom had failed,
By their prowess, to achieve,
They, the children, could and must!

Little thought the Hermit, preaching
Holy Wars to knight and baron,
That the words dropped in his teaching,
His entreaty, his beseeching,
Would by children's hands be gleaned,
And the staff on which he leaned
Blossom like the rod of Aaron.

As a summer wind upheaves
The innumerable leaves
In the bosom of a wood,—
Not as separate leaves, but massed
All together by the blast,—
So for evil or for good
His resistless breath upheaved
All at once the many-leaved,
Many-thoughted multitude.

In the tumult of the air
Rock the boughs with all the nests
Cradled on their tossing crests;
By the fervor of his prayer
Troubled hearts were everywhere
Rocked and tossed in human breasts.

For a century, at least,
His prophetic voice had ceased;
But the air was heated still
By his lurid words and will,
As from fires in far-off woods,
In the autumn of the year,
An unwonted fever broods
In the sultry atmosphere.

II.

CONCERT.

In Cologne the bells were ringing,
In Cologne the nuns were singing
Hymns and canticles divine;
Loud the monks sang in their stalls,
And the thronging streets were loud
With the voices of the crowd;—
Underneath the city walls
Silent flowed the river Rhine.

From the gates, that summer day,
Clad in robes of hodden gray,
With the red cross on the breast,
Azure-eyed and golden-haired,
Forth the young crusaders fared;
While above the band devoted
Consecrated banners floated,
Fluttered many a flag and streamer,
And the cross o'er all the rest !
Singing lowly, meekly, slowly,

Give us, give us back the ho - ly
Se - pul - chre of the Re - deem - er,

Give us, give us.... back the ho - ly

Se - pul - chre of the Re - deem - er.

CONCERT.

On the vast procession pressed,
Youths and maidens. . . .

III.

CONCERT.

Ah! what master hand shall paint
How they journeyed on their way,
How the days grew long and dreary,
How their little feet grew weary,
How their little hearts grew faint!

Ever swifter day by day
Flowed the homeward river; ever
More and more its whitening current
Broke and scattered into spray,
Till the calmly-flowing river
Changed into a mountain torrent,
Rushing from its glacier green
Down through chasm and black ravine.
Like a phœnix in its nest,
Burned the red sun in the West,
Sinking in an ashen cloud;
In the East, above the crest

Of the sea-like mountain chain,
Like a phœnix from its shroud,
Came the red sun back again.

Now around them, white with snow,
Closed the mountain peaks.   Below,
Headlong from the precipice
Down into the dark abyss,
Plunged the cataract white with foam;
And it said, or seemed to say:

SOLO. *f*

Oh   re - turn   while   yet   you

may,....   Fool - ish   chil - dren   to   your

home,   There   the   Ho - ly   Ci - ty

p   *Repeat Chorus.*

is,....   There the Ho - ly   Ci - ty   is.

CONCERT.

But the dauntless leader said:

"Faint not, though your bleeding feet
O'er these slippery paths of sleet
Move but painfully and slowly;
Other feet than yours have bled;
Other tears than yours been shed.
Courage! lose not heart or hope;
On the mountain's southern slope
Lies Jerusalem the Holy!"

CONCERT.

As a white rose in its pride,
By the wind in summer-tide
Tossed and loosened from the branch,
Showers its petals' o'er the ground,
From the distant mountain's side,
Scattering all its snows around,
With mysterious, muffled sound,
Loosened, fell the avalanche.
Voices, echoes far and near,
Roar of winds and waters blending,
Mists uprising, clouds impending,
Filled them with a sense of fear,
Formless, nameless, never ending.

Repeat Chorus, "Give us back," etc., ending thus:

Se - pul - chre... of the Re - deem - er.

Longfellow.

HELIOTROPE.

AMID the chapel's chequered gloom
   She laughed with Dora and with Flora,
And chattered in the lecture-room—
   The saucy little Sophomora !
Yet while (as in her other schools)
   She was a privileged transgressor,
She never broke the simple rules
   Of one particular professor.

But when he spoke of varied lore,
   Paroxytones and moods potential,
She listened with a face that wore
   A look half fond, half reverential.
To her that earnest voice was sweet,
   And, though her love had no confessor,
Her girlish heart lay at the feet
   Of that particular professor.

And he had learned, among his books,
   That held the lore of ages olden,
To watch those ever-changing looks,
   The wistful eyes, and tresses golden,
That stirred his pulse with passion's pain
   And thrilled his soul with soft desire,
Longing for youth to come again,
   Crowned with its coronet of fire.

Her sunny smiles, her winsome ways,
   Were more to him than all his knowledge,
And she preferred his words of praise
   To all honors of the college.

Yet "What am foolish I to him?"
　　She whispered to her one confessor.
"She thinks me old, and gray, and grim,"
　　In silence pondered the professor.

Yet once, when Christmas bells were rung
　　Above ten thousand solemn churches,
And swelling anthems, grandly sung,
　　Pealed through the dim cathedral arches—
Ere home returning, filled with hope,
　　Softly she stole by gate and gable,
And a sweet spray of heliotrope
　　Left on his littered study-table.

Nor came she more, from day to day,
　　Like sunshine through the shadows rifting;
Above her grave, far, far away,
　　The ever-silent snows were drifting:
And those who mourned her winsome face,
　　Found in its stead a swift successor,
And loved another in her place—
　　All, save the silent, old professor.

But in the tender twilight gray,
　　Shut from the sight of carping critic,
His lonely thoughts would often stray
　　From Vedic verse and tongues Semitic—
Bidding the ghost of perished hope
　　Mock with its past the sad possessor
Of the dead spray of heliotrope
　　That once she gave the old professor.
　　　　　　　　From " Acta Columbiana."

THE LAST RIDE.

HIGH o'er the snow-capped peaks of blue the
 stars are out to-night,
And the silver crescent moon hangs low. I
 watched it on my right,
Moving above the pine-tops tall, a bright and
 gentle shape,
While I listened to the tales you told of peril and
 escape.

Then, mingled with your voices low, I heard the
 rumbling sound
Of wheels adown the farther slope, that sought
 the level ground ;
And suddenly, from memories that never can
 grow dim,
Flashed out once more the day when last I rode
 with English Jem.

'Twas here, in wild Montana, I took my hero's
 gauge.
From Butte to Deer Lodge, four-in-hand, he
 drove the mountain stage ;
And many a time, in sun or storm, safe mounted
 at his side,
I whiled away with pleasant talk the long day's
 weary ride.

Jem's faithful steeds had served him long, of
 mettle true and tried :
One sought in vain for trace of blows upon their
 glossy hide ;

And to each low command he spoke the leader's
     nervous ear
Bent eager, as a lover waits his mistress' voice to
     hear.

With ringing crack the leathern whip, that else
     had idly hung,
Kept time for many a rapid mile to English songs
     he sung ;
And yet, despite his smile, he seemed a lonely
     man to be,
With not one soul to claim him kin on this side
     of the sea.

But after I had known him long, one mellow even-
     ing-time
He told me of his English Rose, who withered in
     her prime ;
And how, within the churchyard green, he laid
     her down to rest.
With her sweet babe, a blighted bud, upon her
     frozen breast.

" I could not stay," he said, " where she had left
     me all alone !
The very hedge-rose that she loved I could not
     look upon !
I could not hear the mavis sing, or see the long
     grass wave,
And every little daisy-bank seemed but my dar-
     ling's grave !

" Yet somehow—why, I cannot tell—but when I
     wandered here,

I seemed to bring her with me too, that once had
    been so dear.
I love these mountain summits, where the world
    is in the sky,
For she is in it too,—my love!—and so I bring
    her nigh."

Next week I rode with Jem again. The coach
    was full, that day,
And there were little children there, that pleased
    us with their play.
A sweet-faced mother brought her pair of rosy,
    bright-eyed girls,
And boy like one I left at home, with silken yel-
    low curls.

We took fresh horses at Girard's, and as he led
    them out—
A vicious pair they seemed to me—I heard the
    hostler shout:
" You always want good horses, Jem! Now you
    shall have your way.
Try these new beauties, for we sold your old team
    yesterday."

O'er clean-cut limb and sloping flank, arched
    neck and tossing head,
I marked Jem run his practised eye, though not
    a word he said;
Yet, as he clambered to his seat, and took the
    reins once more,
I saw a look upon his face it had not worn
    before.

The hostler open flung the gates. "Now, Tempest, show your pace,"
He cried, and with a careless hand he struck the leader's face.
The horse, beneath the sportive blow, reared as if poison-stung;
And, with his panic-stricken mates, to a mad gallop sprung.

We thundered through the gate, and out upon the stony road ;
From side to side the great coach lurched, with all its priceless load :
Some cried aloud for help, and some, with terror-frozen tongue,
Clung, bruised and faint in every limb, the weaker to the strong.

And men who oft had looked on death, unblanched, by flood or field,
When every nerve to do and dare by agony was steeled,
Now moaned aloud, or gnashed their teeth in helpless rage,
To die, at whim of maddened brutes, like vermin in a cage !

Too well, alas! too well I knew the awful way we went,—
The little stretch of level road, and then the steep descent ;
The boiling stream that seethed and roared far down the rocky ridge,

With death, like old Horatius, grim waiting at
    the bridge!

But, suddenly, above the din, a voice rang loud
    and clear ;
We knew it well, the driver's voice,—without one
    note of fear ;
Some strong, swift angel's lips might thrill with
    such a clarion cry,—
The voice of one who put for aye all earthly pas-
    sion by :—

"Still! for your lives, and listen! See yon farm-
    house by the way,
And piled along the field in front the shocks of
    new-mown hay.
God help me turn my horses there! And when
    I give the word,
Leap on the hay! Pray, every soul, to Him who
    Israel heard!"

Within, the coach was still. 'Tis strange, but
    never till I die
Shall I forget the fields that day, the color of the
    sky,
The summer breeze that brought the first sweet
    perfume of the hay,
The bobolink that in the grass would sing his
    life away.

One breathless moment bridged the space that
    lay between, and then
Jem drew upon the straining reins, with all the
    strength of ten.

"Hold fast the babes!" More close I clasped
    the fair boy at my side.
"Let every nerve be steady now!" and "Jump
    for life!" he cried.

Saved, every soul! Oh! dizzy—sweet life rushed
    in every vein,
To us who from that fragrant bed rose up to hope
    again!
But, 'mid the smiles and grateful tears that
    mingled on each cheek,
A sudden questioning horror grew, that none
    would dare to speak.

Too soon the answer struck our ears! One mo-
    ment's hollow roar
Of flying hoofs upon the bridge—an awful crash
    that tore
The very air in twain—and then, through all the
    world grown still,
I only heard the bobolink go singing at his will.

I was the first man down the cliff. There's little
    left to tell.
We found him lying, breathing yet and conscious,
    where he fell.
The question in his eager eyes, I answered with
    a word,—
"Safe!" Then he smiled, and whispered low
    some words I scarcely heard.

We would have raised him, but his lips grew
    white with agony.

"Not yet; it will be over soon," he whispered.
    " Wait with me ;"
Then, lower, smiling still, "It is my last ride,
    friends; but I
Have done my duty, and God knows I do not
    fear to die."

He closed his eyes. We watched his life slip,
    like an ebbing tide,
Far out upon the infinite, where all our hopes
    abide.
He spoke but once again, a name not meant for
    mortal ears,—
"My Rose!" She must have heard that call,
    amid the singing spheres!
                  Mary A. P. Stansbury.

THE GALLEY SLAVE.

There lived in France, in days not long now dead,
    A farmer's sons, twin brothers, like in face ;
And one was taken in the other's stead
    For a small theft, and sentenced in disgrace
To serve for years a hated galley slave—
Yet said no word, his prized good name to save.

Trusting remoter days would be more blessed,
    He set his will to wear the verdict out,
And knew most men are prisoners at best,
    Who some strong habit ever drag about,
Like chain and ball; then meekly prayed that he
Rather the prisoner he was should be.

But best resolves are of such feeble thread,
  They may be broken in temptation's hands.
After long toil the guiltless prisoner said:
  Why should I thus, and feel life's precious sands
The narrow of my glass, the present, run,
For a poor crime that I have never done?

Such questions are like cups, and hold reply;
  For when the chance swung wide the prisoner
      fled,
And gained the country road, and hastened by
  Brown furrowed fields and skipping brooklets,
      fed
By shepherd clouds, and felt 'neath sapful trees
The soft hand of the mesmerizing breeze.

Then, all that long day having eaten naught,
  He at a cottage stopped, and of the wife
A brimming bowl of fragrant milk besought.
  She gave it him; but as he quaffed the life,
Down her kind face he saw a single tear
Pursue its wet and sorrowful career.

Within the cot he now beheld a man
  And maiden, also weeping. "Speak," said he,
"And tell me of your grief; for if I can,
  I will disroot the sad, tear-fruited tree."
The cotter answered: "In default of rent,
We shall to-morrow from this roof be sent."

Then said the galley slave: "Whoso returns
  A prisoner escaped may feel the spur
To a right action, and deserves and earns
  Proffered reward. I am a prisoner!

Bind these my arms, and drive me back my way,
That your reward the price of home may pay."

Against his wish the cotter gave consent,
   And at the prison-gate received his fee,
Though some made it a thing for wonderment
   That one so sickly and infirm as he,
When stronger would have dared not to attack,
Could capture this bold youth and bring him
      back.

Straightway the cotter to the mayor hied,
   And told him all the story ; and that lord
Was much affected, dropping gold beside
   The pursed sufficient silver of reward ;
Then wrote his letter in authority,
Asking to set the noble prisoner free.

There is no nobler, better life on earth
   Than that of conscious, meek self-sacrifice.
Such life our Saviour, in His lowly birth
   And holy work, made His sublime disguise—
Teaching this truth, still rarely understood :
'Tis sweet to suffer for another's good.
                 Henry Abbey.

THE SEA BREEZE AND THE SCARF.

HUNG on the casement that looked o'er the main
   Fluttered a scarf of blue ;
And a gay, bold breeze paused to flutter and
   tease
   This trifle of delicate hue.

" You are lovelier far than the proud skies are,"
  He said, with a voice that sighed.
" You are fairer to me than the beautiful sea;
  Oh! why do you stay here and hide?

" You are wasting your life in this dull, dark
    room;"
  And he fondled her silken folds.
" O'er the casement lean but a little, my queen,
  And see what the great world holds.
How the wonderful blue of your matchless hue
  Cheapens both sea and sky!
You are far too bright to be hidden from sight:
  Come, fly with me, darling, fly!"

Tender his whisper, and sweet his caress;
  Flattered and pleased was she:
The arms of her lover lifted her over
  The casement out to sea.
Close to his breast she was fondly pressed,
  Kissed once by his laughing mouth:
Then dropped to her grave in the cruel wave,
  While the wind went whistling south.
              Ella Wheeler Wilcox.

THE MANTLE OF ST. JOHN DE MATHA.

A STRONG and mighty Angel,
  Calm, terrible, and bright,
The cross in blended red and blue
  Upon his mantle white!

Two captives by him kneeling,
  Each on his broken chain,

Sang praise to God who raiseth
The dead to life again !

Dropping his cross-wrought mantle,
" Wear this," the Angel said ;
"Take thou, O Freedom's priest, its sign,—
The white, the blue, and red."

Then rose up John de Matha
In the strength the Lord Christ gave,
And begged through all the land of France
The ransom of the slave.

The gates of tower and castle
Before him open flew,
The drawbridge at his coming fell,
The door-bolt backward drew.

For all men owned his errand,
And paid his righteous tax ;
And the hearts of lord and peasant
Were in his hands as wax.

At last, outbound from Tunis,
His bark her anchor weighed :
Freighted with seven-score Christian souls
Whose ransom he had paid.

But, torn by Paynim hatred,
Her sails in tatters hung ;
And on the wild waves, rudderless,
A shattered hulk she swung.

" God save us !" cried the captain,
 For naught can man avail ;
Oh, woe betide the ship that lacks
 Her rudder and her sail !

" Behind us are the Moormen ;
 At sea we sink or strand :
There's death upon the water,
 There's death upon the land !"

Then up spake John de Matha :
 " God's errands never fail !
Take thou the mantle which I wear,
 And make of it a sail."

They raised the cross-wrought mantle,
 The blue, the white, the red ;
And straight before the wind off-shore
 The ship of Freedom sped.

" God help us !" cried the seamen,
 " For vain is mortal skill :
The good ship on a stormy sea
 Is drifting at its will."

Then up spake John de Matha :
 " My mariners, never fear !
The Lord whose breath has filled her sail
 May well our vessel steer !"

So on through storm and darkness
 They drove for weary hours ;

And lo ! the third gray morning shone
  On Ostia's friendly towers.

And on the walls the watchers
  The ship of mercy knew,—
They knew far off its holy cross,
  The red, the white, and blue.

And the bells in all the steeples
  Rang out in glad accord,,
To welcome home to Christian soil
  The ransomed of the Lord.

Whittier.

SLEEP.

"He giveth His belovèd sleep."—*Ps.* cxxvii. 2.

Of all the thoughts of God that are
Borne inward into souls afar
Along the Psalmist's music deep,
Now tell me if that any is,
For gift or grace, surpassing this,—
" He giveth His belovèd sleep."

What would we give to our beloved?
The hero's heart to be unmoved,
The poet's star-tuned harp to sweep,
The patriot's voice to teach and rouse,
The monarch's crown to light the brows?—
He giveth His belovèd sleep.

What do we give to our beloved?
A little faith all undisproved,
A little dust to overweep,

And bitter memories to make
The whole earth blasted for our sake:
He giveth His belovèd sleep.

"Sleep soft, beloved!" we sometimes say,
Who have no tune to charm away
Sad dreams that through the eyelids creep;
But never doleful dream again
Shall break the happy slumber when
He giveth His belovèd sleep.

O earth, so full of dreary noises!
O men with wailing in your voices!
O delvèd gold the wailers heap!
O strife, O curse, that o'er it fall!
God strikes a silence through you all,
And giveth His belovèd sleep.

His dews drop mutely on the hill,
His cloud above it saileth still,
Though on its slope men sow and reap:
More softly than the dew is shed,
Or cloud is floated overhead,
He giveth His belovèd sleep.

Ay, men may wonder while they scan
A living, thinking, feeling man
Confirmed in such a rest to keep;
But angels say, and through the word
I think their happy smile is *heard*,
"He giveth His belovèd sleep."

For me, my heart that erst did go
Most like a tired child at a show,
That sees through tears the mummers leap,
Would now its wearied vision close,
Would childlike on His love repose
Who giveth His belovèd sleep.

And friends, dear friends, when it shall be
That this low breath has gone from me,
And round my bier ye come to weep,
Let one most loving of you all
Say, " Not a tear must o'er her fall !
He giveth His belovèd sleep."

Mrs. Browning.

THE LEGEND OF ST. MARK.

THE day is closing dark and cold,
   With roaring blast and sleety showers ;
And through the dusk the lilacs wear
   The bloom of snow, instead of flowers.

I turn me from the gloom without,
   To ponder o'er a tale of old ;
A legend of the age of Faith,
   By dreaming monk or abbess told.

On Tintoretto's canvas lives
   That fancy of a loving heart,
In graceful lines and shapes of power,
   And hues immortal as his art.

In Provence (so the story runs)
  There lived a lord, to whom, as slave,
A peasant boy of tender years
  The chance of trade or conquest gave.

Forth-looking from the castle tower,
  Beyond the hills with almonds dark,
The straining eye could scarce discern
  The chapel of the good St. Mark.

And there, when bitter word or fare
  The service of the youth repaid,
By stealth, before that holy shrine,
  For grace to bear his wrong, he prayed.

The steed stamped at the castle gate,
  The boar-hunt sounded on the hill ;
Why stayed the Baron from the chase,
  With looks so stern, and words so ill?

"Go, bind yon slave! and let him learn,
  By scath of fire, and strain of cord,
How ill they speed who give dead saints
  The homage due their living lord!"

They bound him on the fearful rack,
  When, through the dungeon's vaulted dark,
He saw the light of shining robes,
  And knew the face of good St. Mark.

Then sank the iron rack apart,
  The cords released their cruel clasp,
The pincers, with their teeth of fire,
  Fell broken from the torturer's grasp.

And lo ! before the youth and saint,
   Barred door and wall of stone gave way ;
And up from bondage and the night
   They passed to freedom and the day !

Whittier.

SCENES FROM KING RICHARD III.

Act III., Scene I.

The Hall in Crosby Palace.

DUKE OF BUCKINGHAM, DUKE OF GLOSTER, and PRINCE
EDWARD.

Buck. Welcome, sweet prince, to London, to
   your chamber.
Glos. Welcome, dear cousin, my thoughts' sov-
   ereign :
The weary way hath made you melancholy.
   Prince E. No, uncle ; but our crosses on the
      way
Have made it tedious, wearisome and heavy :
I want more uncles here to welcome me.
   Glos. Sweet prince, the untainted virtue of
      your years
Hath not yet div'd into the world's deceit ;
Nor more can you distinguish of a man
Than of his outward show ; which, God he knows,
Seldom or never jumpeth with the heart.
Those uncles which you want were dangerous ;
Your grace attended to their sugar'd words,
But look'd not on the poison of their hearts ;
God keep you from them, and from such false
      friends !

Prince E. God keep me from false friends! but they were none.

Buck. My lord, the Mayor of London comes to greet you.

Enter the LORD MAYOR.

Lord Mayor. God bless your grace with health and happy days!

Prince E. I thank you, good my lord; and thank you all.
I thought my mother, and my brother York,
Would long ere this have met us on the way;
Say, uncle Gloster, if our brother come,
Where shall we sojourn till our coronation?

Glos. Where it seems best unto your royal self.
If I may counsel you, some day or two
Your highness shall repose you at the Tower;
Then where you please and shall be thought most fit
For your best health and recreation.

Prince E. I do not like the Tower, of any place.
Did Julius Cæsar build that place, my lord?

Buck. He did, my gracious lord, begin that place;
Which, since, succeeding ages have re-edified.

Prince E. Is it upon record, or else reported
Successively from age to age, he built it?

Buck. Upon record, my gracious lord.

Prince E. But say, my lord, it were not register'd.
Methinks the truth should live from age to age,

As 'twere retail'd to all posterity,
Even to the general all-ending day.

Glos. So wise so young, they say, do ne'er live
long.

Prince E. What say you, uncle?

Glos. I say, without characters, fame lives
long.

Thus, like the formal vice, Iniquity,
I moralize two meanings in one word.

Prince E. That Julius Cæsar was a famous
man;

With what his valor did enrich his wit,
His wit set down to make his valor live:
Death makes no conquest of this conqueror;
For now he lives in fame, though not in life.
I'll tell you what, my cousin Buckingham,—

Buck. What, my gracious lord?

Prince E. An if I live until I be a man,
I'd win our ancient right in France again,
Or die a soldier, as I lived a king.

Glos. Short summers lightly have a forward
spring.

Enter DUKE OF YORK *and* HASTINGS.

Buck. Now, in good time, here comes the Duke
of York.

Prince E. Richard of York! how fares our loving brother?

Duke Y. Well, my dread lord; so must I call
you now.

Prince E. Ay, brother, to our grief, as it is
yours:

Too late he died that might have kept that title,
Which by his death hath lost much majesty.

   Glos. How fares our cousin, noble lord of
     York?

   Duke Y. I thank you, gentle uncle. O, my
     lord,

You said that idle weeds are fast in growth:
The prince my brother hath outgrown me far.

   Glos. He hath, my lord.

   Duke Y. And therefore is he idle?

   Glos. O, my fair cousin, I must not say so.

   Duke Y. Then is he more beholding to you
     than I.

   Glos. He may command me as my sovereign;

But you have power in me as in a kinsman.

   Duke Y. I pray you, uncle, give me this dagger.

   Glos. My dagger, little cousin? with all my
     heart.

   Prince E. A beggar, brother?

   Duke Y. Of my kind uncle, that I know will
     give;

And being but a toy, which is no grief to give.

   Glos. A greater gift than that I'll give my
     cousin.

   Duke Y. A greater gift! O, that's the sword
     to it.

   Glos. Ay, gentle cousin, were it light enough.

   Duke Y. O, then, I see, you will part but with
     light gifts;

In weightier things you'll say a beggar nay.

   Glos. It is too heavy for your grace to wear.

   Duke Y. I weigh it lightly, were it heavier.

Glos. What, would you have my weapon, little
lord ?

Duke Y. I would, that I might thank you as
you call me.

Glos. How ?

Duke Y. Little.

Prince E. My Lord of York will still be cross
in talk :

Uncle, your grace knows how to bear with him.

Duke Y. You mean, to bear me, not to bear
with me :

Uncle, my brother mocks both you and me ;

Because that I am little, like an ape,

He thinks that you should bear me on your
shoulders.

Buck. With what a sharp-provided wit he
reasons !

To mitigate the scorn he gives his uncle,

He prettily and aptly taunts himself :

So cunning and so young is wonderful.

Glos. My lord, will't please you pass along ?

Myself and my good cousin Buckingham

Will to your mother, to entreat of her

To meet you at the Tower and welcome you.

Duke Y. What! will you go unto the Tower,
my lord ?

Prince E. My lord protector needs will have
it so.

Duke Y. I shall not sleep in quiet at the Tower.

Glos. I'll warrant you :—King Henry lay there,

And he sleeps in quiet.

Why, what should you fear ?

Duke Y. Marry, my uncle Clarence's angry
    ghost:
My grandam told me he was murder'd there.
    Prince E. I fear no uncles dead.
    Glos. Nor none that live, I hope.
    Prince E. An if they live, I hope I need not
    fear.
But come, my lord; and with a heavy heart,
Thinking on them, go I unto the Tower.

Act IV., Scene I.

Within the Tower.

LADY ANNE, DUKE OF YORK, QUEEN, PRINCE EDWARD, and
DUCHESS OF YORK.

    Prince E. Pray, madam, do not leave me yet,
For I have many more complaints to tell you.
    Queen. And I unable to redress the least;
What wouldst thou say, my child?
    Prince E. Oh, mother, since I have lain i' the
    Tower
My rest has still been broke with frightful dreams,
Or shocking news has wak'd me into tears;
I'm scarce allow'd a friend to visit me;
All my old honest servants are turn'd off,
And in their room are strange ill-natur'd fellows,
Who look so bold, as they were all my masters;
And I'm afraid they'll shortly take you from me.
    Duc. Y. Oh, mournful hearing!
    Lady A. Oh, unhappy prince!
    Duke Y. Dear brother, why do you weep so?
You make me cry too.
    Queen. Alas, poor innocent!

Prince E. Would I but knew at what my uncle
    aims ;
If 'twere my crown, I'd freely give it him,
So that he'd let me joy my life in quiet.
    Duke Y. Why, will my uncle kill us, brother ?
    Prince E. I hope he won't ; we never injured
    him.
    Queen. I cannot bear to see 'em thus.

        Enter LORD STANLEY.

    Lord Stan. Come, madam, you must straight
    to Westminster,
There to be crowned Richard's royal queen.
    Anne. Despiteful tidings ! O unpleasing news !
    Stan. Come, madam, come ; I in all haste was
    sent.
    Anne. And I in all unwillingness will go.
I would to God that the inclusive verge
Of golden metal that must round my brow
Were red-hot steel, to sear me to the brain !
Anointed let me be with deadly venom,
And die ere men can say, God save the queen !
    Queen. Go, go, poor soul, I envy not thy glory ;
To feed my humor, wish thyself no harm.
    Anne. No ! why? When he that is my hus-
    band now
Came to me, as I follow'd Henry's corse,
When scarce the blood was well washed from his
    hands
Which issued from my other angel husband,
And that dead saint which then I weeping fol-
    low'd ;

O, when, I say, I look'd on Richard's face,
This was my wish: "Be thóu," quoth I, " ac-
    curs'd,
For making me, so young, so old a widow !
And, when thou wed'st, let sorrow haunt thy
    bed ;
And be thy wife—if any be so mad—
As miserable by the life of thee
As thou hast made me by my dear lord's death !"
Lo, ere I can repeat this curse again,
Even in so short a space, my woman's heart
Grossly grew captive to his honey words
And proved the subject of my own soul's curse,
Which ever since hath kept my eyes from rest ;
For never yet one hour in his bed
Have I enjoy'd the golden dew of sleep,
But have been waked by his timorous dreams.
Besides, he hates me for my father Warwick :
And will, no doubt, shortly be rid of me.

    Queen. Poor heart, adieu! I pity thy com-
        plaining.
    Anne. No more than from my soul I mourn for
        yours.
    Queen. Farewell, thou woful welcomer of
        glory !
    Anne. Adieu, poor soul, that takest thy leave
        of it !
    Duc. of Y. Go thou to Richard, and good angels
        guard thee.
    [*To Queen.*] Go thou to sanctuary, and good
        thoughts possess thee !
I, to my grave, where peace and rest lie with me !

Eighty odd years of sorrow have I seen,
And each hour's joy wreck'd with a week of teen.
   Prince E. Dear madam, take me hence: for I
   shall ne'er
Enjoy a moment's quiet here.
   Duke Y. Nor I; pray, mother, let me go too.
   Queen. Come, then, my pretty young ones, let's
   away;
For here you lie within the falcon's reach,
Who watches but th' unguarded hour to seize
   you.

Enter Lieutenant, L.

   Lieut. I beg your majesty will pardon me:
But the young princes must on no account
Have egress from the Tower:
Nor must (without the king's especial license),
Of what degree soever, any person
Have admittance to 'em :—all must retire.
   Queen. I am their mother, sir; who else com-
   mands 'em?
If I pass freely they shall follow me.
For you, I'll take the peril of your fault upon
   myself.
   Lieut. My inclination, madam, would oblige
   you;
But I am bound by oath, and must obey:
Nor, madam, can I now with safety answer
For this continued visit.
   Queen. Oh, heavenly powers! shall not I stay
   with them?
   Lieut. Such are the king's commands, madam.

Queen. Support me, heaven !
For life can never bear the pangs of such a
    parting.
Oh, my poor children ! Oh, distracting thought !
I dare not bid 'em as I should, farewell ;
And then to part in silence stabs my soul !
    Prince E. What, must you leave us, mother ?
    Queen. What shall I say ?
But for a time, my loves :—we shall meet again :
At least in heaven.
    Duke Y. Won't you take me with you, mother ?
I shall be so 'fraid to stay, when you are gone.
    Queen. I cannot speak to 'em, and yet we must
Be parted.
Then let these kisses say farewell.
Why, oh why, just heaven, must these be our
    last !
    Duc. Y. Give not your grief such way :—be
        sudden when you part.
    Queen. I will : since it must be so :—to heaven
        I leave 'em.
Hear me, ye guardian powers of innocence !
Awake or sleeping, oh protect 'em still !
Still may their helpless youth attract men's pity,
That when the arm of cruelty is raised,
Their looks may drop the lifted dagger down
From the stern murderer's relenting hand,
And throw him on his knees in penitence !
    Prince E. ⎫
    Duke Y. ⎬ Oh, mother ! mother !

    Queen.   Oh, my poor children !
                        Shakespeare (Adapted).

MOTHER AND POET.

Turin, after News from Gaeta, 1861.

DEAD! One of them shot by the sea in the east,
    And one of them shot in the west by the sea.
Dead! both my boys! When you sit at the
    feast
    And are wanting a great song for Italy free,
      Let none look at *me!*

Yet I was a poetess only last year,
    And good at my art, for a woman, men said;
But *this* woman, *this,* who is agonized here,
    —The east sea and west sea rhyme on in her
      head
      Forever instead.

What art's for a woman? To hold on her knees
    Both darlings! to feel all their arms round her
      throat,
Cling, strangle a little! to sew by degrees
    And 'broider the long-clothes and neat little
      coat;
      To dream and to dote.

To teach them. . . . It stings there! *I* made
    them indeed
    Speak plain the word *country.* *I* taught them,
      no doubt,
That a country's a thing men should die for at
    need.
    I prated of liberty, rights, and about
      The tyrant cast out.

And when their eyes flashed . . . O my beautiful
    eyes! . . .
  I exulted; nay, let them go forth at the wheels
Of the guns, and denied not. But then the sur-
    prise
  When one sits quite alone! Then one weeps,
    then one kneels.
    God, how the house feels!

At first, happy news came, in gay letters moiled
  With my kisses,—of camp-life and glory, and
    how
They both loved me; and, soon coming home to
    be spoiled,
  In return would fan off every fly from my brow
    With their green laurel-bough.

Then was triumph at Turin: "Ancona was free!"
  And some one came out of the cheers in the
    street,
With a face pale as stone, to say something to
    me.
  My Guido was dead! I fell down at his feet,
    While they cheered in the street.

I bore it; friends soothed me; my grief looked
    sublime
  As the ransom of Italy. One boy remained
To be leant on and walked with, recalling the
    time
  When the first grew immortal, while both of us
    strained
    To the height he had gained.

And letters still came, shorter, sadder, more
    strong,
    Writ now but in one hand, "I was not to
      faint,—
One loved me for two—would be with me ere long:
    And *Viva l'Italia !—he* died for, our saint,
      Who forbids our complaint."

My Nanni would add, " he was safe, and aware
    Of a presence that turned off the balls,—was
      imprest
It was Guido himself, who knew what I could
    bear,
    And how 'twas impossible, quite dispossest,
      To live on for the rest."

On which, without pause, up the telegraph-line
    Swept smoothly the next news from Gaeta :—
      Shot.
Tell his mother. Ah, ah! "his," "their" mother,—
    not "mine,"
    No voice says "*My* mother" again to me.
      What!
      You think Guido forgot?

Are souls straight so happy, that, dizzy with
    heaven,
    They drop earth's affections, conceive not of
      woe ?
I think not. Themselves were too lately for-
    given

Through THAT Love and Sorrow which recon-
 ciled so
  The Above and Below.

Both boys dead? but that's out of nature. We
 all
  Have been patriots, yet each house must al-
   ways keep one.
'Twere imbecile, hewing out roads to a wall;
  And when Italy's made, for what end is it
   done
   If we have not a son?

Ah, ah, ah! when Gaeta's taken, what then?
  When the fair wicked queen sits no more at her
   sport
Of the fire-balls of death crashing souls out of
 men ?
  When the guns of Cavalli with final retort
   Have cut the game short?

When Venice and Rome keep their new jubilee,
  When your flag takes all heaven for its white,
   green, and red,
When *you* have your country from mountain to ·
 sea,
  When King Victor has Italy's crown on his
   head,
   (And *I* have my Dead):—

What then? Do not mock me. Ah, ring your
 bells low,
  And burn your lights faintly! *My* country is
   there,

Above the star pricked by the last peak of snow :
　My Italy's THERE, with my brave civic Pair
　　To disfranchise despair !

Dead ! One of them shot by the sea in the east,
　And one of them shot in the west by the sea.
Both ! both my boys ! If in keeping the feast
　You want a great song for your Italy free,
　　Let none look at *me !*

[This was Laura Savio, of Turin, a poetess and patriot,
whose sons were killed at Ancona and Gaeta.]

Mrs. Browning.

SCENE IN A TENEMENT HOUSE.

I WENDED my way through wind and snow
One winter's night to tenement row ;
The place seemed under the ban and blight
Of a ghostly spell, that stormy night.
Unearthly footsteps seemed to fall
In the dismal darkness down the hall ;
Unearthly voices, deep and low,
Seemed to whisper a tale of woe.

From reeking angle, and rotten stair,
As through the foul and fetid air
I groped along, to a broken door
Of a certain room, or rather den,
Such as some wealthy, prosperous men
Build, and rent to the homeless poor.
The door was ajar : within all dark,
Never an ember, never a spark

Glowed or glimmered athwart the gloom
That hung like a pall in that wretched room.

But I heard the patter of children's feet,
And sounds of voices low and sweet;
And one, he was only three years old,
Said : "Sister, wot makes mamma so told ?
Pease et me ake her," the sweet voice plead.
"I's so hungry, I 'onts some bread.
On'y the 'ittlest piece 'ill do,
And Johnnie 'ill give a bit to you."

"Hush, Johnnie, hush," the sister said ;
"There's not a single crust of bread.
Don't wake poor mamma, she's sick, you know,
So sick and weak, she cannot sew.
Don't you remember how she cried,
When she bade me put my work aside,
And how she kissed us when she said,
The Father in Heaven will give us bread ?

"All day long through snow and sleet
I wandered up and down the street,
And, Johnnie, I held my freezing hand
To crowds of ladies rich and grand.
But they did not hear me when I said,
'Please give me a penny to buy some bread ;'
One beautiful lady turned and smiled,
But she only said, 'Don't touch me, child.'

"In their splendid clothes they all swept by,
And I was so cold, but I did not cry.

Oh ! Johnnie, I never begged before,
But I went to-day from door to door,
Till my very heart grew faint and weak,
And I shivered so, I could not speak.
But when I remembered what mamma said :
' The Father in Heaven will give you bread,'
I forgot all the hunger and pain,
And went on asking, and asking in vain,
'Till I could hardly move my freezing feet.
And when they lighted the lamps in the street,
I came away through mud and mire,
With nothing to eat or to make a fire ;
But as I was passing Denny's shop,
Some one called out, ' Stop, Katie, stop !'
And out came little Sammie Dole,
And filled my basket with wood and coal,
So now we can have a fire, you see,
And oh ! how nice and warm it will be ;
And, Johnnie, if you will be still and good,
I'll tell you Little Red Riding Hood."

" No, no, I's hungry," the wee one said ;
" Tant oo dive me a 'ittle bread ?
Dest a trum, I sint oo tood,
And Johnnie 'ill do to seep and be dood."

" There's not a crumb of bread ; don't cry,
In the morning sister will try
To get poor mamma a bit of meat
And some nice bread for Johnnie to eat."

By this time, the little cold blue hands
Had heaped together some half charred brands

And kindled a fire. Oh! surely, the light
Never revealed a sadder sight
Than greeted my eyes that winter night.
Walls damp and broken, a window bare,
A rickety table, a bottomless chair,
A floor discolored by soil and stain,
Snow driving in through the window pane.

Wee womanly Katie, scarce nine years old,
Pinched and shrunken from hunger and cold,
Sweet baby Johnnie, with dimpled feet,
Crying and pleading for something to eat,
A tattered bed, where the eye could trace
A human form, with a sad white face,
A sad white face, that had once been fair,
Framed in a tangle of light brown hair;
The sad eyes closed, the lips apart,
Small white hands crossed on a quiet heart.

Softly Katie approached her now,
And pressed a kiss on that marble brow,
Then with a smothered cry she said:
"Johnnie, oh! Johnnie, mamma is dead!"
Speak to me, mamma, one word!" she cried,
" O speak to your Katie!" No voice replied.

But Johnnie crept to the quiet breast
Where the golden head was wont to rest,
And nestling close to the icy form,
Said, " I tan teep sweet mamma warm."

But the mother, outworn in the struggle and
        strife
Of the madness and toil of the battle of life,

Had silently gone to that beautiful shore
Where the rich hath need of their gold no
   more.

THE FEMALE MARTYR.

"BRING out your dead!" The midnight street
  Heard and gave back the hoarse, low call;
Harsh fell the tread of hasty feet,
Glanced through the dark the coarse white
   sheet,
  Her coffin and her pall.
"What—only one!" the brutal hack-man said,
As with an oath he spurned away the dead.

How sunk the inmost hearts of all,
  As rolled that dead-cart slowly by,
With creaking wheel and harsh hoof-fall!
The dying turned him to the wall,
  To hear it and to die!
Onward it rolled; while oft its driver stayed,
And hoarsely clamored, "Ho! bring out your
   dead."

It paused beside the burial-place;
  "Toss in your load!"—and it was done.
With quick hand and averted face,
Hastily to the grave's embrace
  They cast them, one by one,
Stranger and friend, the evil and the just,
Together trodden in the churchyard dust!

And thou, young martyr! thou wast there;
  No white-robed sisters round thee trod,
Nor holy hymn, nor funeral prayer
Rose through the damp and noisome air
  Giving thee to thy God;
Nor flower, nor cross, nor hallowed taper gave
Grace to the dead, and beauty to the grave!

Yet, gentle sufferer! there shall be,
  In every heart of kindly feeling,
A rite as holy paid to thee
As if beneath the convent-tree
  Thy sisterhood were kneeling,
At vesper hours, like sorrowing angels, keeping
Their tearful watch around thy place of sleeping.

For thou wast one in whom the light
  Of Heaven's own love was kindled well;
Enduring with a martyr's might,
Through weary day and wakeful night,
  Far more than words may tell:
Gentle, and meek, and lowly, and unknown,
Thy mercies measured by thy God alone!

Where manly hearts were failing, where
  The throngful street grew foul with death,
O high-souled martyr! thou wast there,
Inhaling, from the loathsome air,
  Poison with every breath,
Yet shrinking not from offices of dread
For the wrung dying, and the unconscious dead.

And, where the sickly taper shed
  Its light through vapors, damp, confined,
Hushed as a seraph's fell thy tread,
A new Electra by the bed
  Of suffering human-kind!
Pointing the spirit, in its dark dismay,
To that pure hope which fadeth not away.

 Innocent teacher of the high
  And holy mysteries of Heaven!
How turned to thee each glazing eye,
In mute and awful sympathy,
  As thy low prayers were given;
And the o'er-hovering Spoiler wore, the while,
An angel's features,—a deliverer's smile!

A blessed task! and worthy one
  Who, turning from the world, as thou,
Before life's pathway had begun
To leave its spring-time flowers and sun,
  Had sealed her early vow;
Giving to God her beauty and her youth,
Her pure affections and her guileless truth.

Earth may not claim thee. Nothing here
  Could be for thee a meet reward;
Thine is a treasure far more dear:
Eye hath not seen it, nor the ear
  Of living mortal heard,
The joys prepared, the promised bliss above,
The holy presence of Eternal Love!

Sleep on in peace.   The earth has not
  A nobler name than thine shall be.
The deeds by martial manhood wrought,
The lofty energies of thought,
  The fire of poesy,
These have but frail and fading honors ; thine
Shall Time unto Eternity consign.

Yea, and when thrones shall crumble down,
  And human pride and grandeur fall,
The herald's line of long renown,
The mitre and the kingly crown,—
  Perishing glories all !
The pure devotion of thy generous heart
Shall live in Heaven, of which it was a part.

Whittier.

PANCRATIUS.

A HUSH lay on the multitudes.   Softly and low
Died out the echoes of that mighty roar,
Which rose triumphant but a space ago,
As the strong wrestler, pale as Alpine snow,
Reeled in his agony, and stirred no more.

They bore him forth, and in his robe of pride
The Roman courtier turned with smiling face,
To woo the fair girl resting at his side,
Who, in her beauty, calm and starry-eyed,
Could view such struggles with a careless grace.

But hark !   Along the smiling, sparkling tier,
A murmur stole—the smile gave place to frown,
And every eager eye grew cold and clear,

When light and graceful as a mountain deer,
A Christian martyr sprang to win his crown!

It was a youth—a slight yet manly form—
Who, with an eye like some unruffled lake,
And virgin cheek with rosy blushes warm,
Seemed all too tender for the cruel storm
Whose giant force must either bend or break.

And yet there was a calm upon the brow,
And in those thoughtful eyes a holy peace
As though the youthful martyr stood e'en now
In triumph on a noble vessel's prow,
Whose port was nigh, whose labors soon should
    cease.

Slowly he turned, and o'er the swaying tide
Of jeweled forms his gentle glance was flung
Till many a Roman maiden turned aside,
Lest some might note the grief she could not hide
At thought of death to one so fair and young.

But pity, like the trembling moonbeam shed
Athwart the dark waves of a stormy sea,
O'er those untutored hearts, by passion led,
Gleamed but a fitful space, then meekly fled,
As things of light from darkness ever flee.

And he, Pancratius, in his joyous race,
Was nearing fast the long desired goal
Ere age had dashed the beauty from that face,
Whose shrine should be in time the fitting place
To nerve the fainting faith or sinking soul!

He stood unmoved, e'en as the warrior stands
Who neither courts nor shuns the coming fray ;
But even as he clasped his slender hands,
A door swung grating—and across the sands
A lion stalked in majesty of might.

There was no fury in his stately tread,
No bloody thirst which hastens to destroy,
But calm in power he raised his noble head,
And with a kingly glory 'round him shed,
Moved onward to that slender, graceful boy.

Nearer he came ; upon the martyr's cheek
The hot breath of the forest-monarch burned,
Till once—but once—that brave young heart grew
    weak,
When lo! with startled look, all mild and meek,
Back to its den the moaning lion turned!

Then rose that mighty multitude and loud
Upswelled a shout of mingled joy and rage,
As some their gladly tearful faces bowed,
While others stood apart and, stormy-browed,
Chafed like the maniac in his iron cage.

But o'er that tide of sound which rudely gushed
Till Tiber all her slumbering echoes woke—
A clear young voice rang out, the din was
    hushed,
And while his brow, uplifted, brightly blushed,
With gentle grace, the young Pancratius spoke:

"Patience, sweet friends," he cried, "bear yet
    awhile,
For see, yon panther thirsts for liberty.
'Twas he that freed my father from his toil;
Oh! may he not"—and here a glorious smile
Parted his bright lips—" set Pancratius free?"

He paused—and men gazed, wonder-stricken, how
Such thirst could be for that which mortals dread;
Yet with a gloomy satisfaction on each brow,
The fatal sign was made, and cageless, now
A panther bounded forth with noiseless tread.

Joyous in liberty, it frisked and played,
And turned its shining neck in conscious pride;
Now in the yielding sand its form was laid;
Anon, with cat-like glee, low murmurs made,
And shook the dusk sand from its glittering hide.

At length it rose—its keen quick glance had
    caught
The youthful martyr, as he stood apart,
With all a mother's tender lips had taught,
And all a Saviour's tender love had wrought,
In that dread moment stealing o'er his heart.

Earnest the Christian prayed, and breathless,
    men
Beheld the look that crouching panther wore;
There was a pause—the echoes slept again—
And then—oh! just and righteous Father! then
One bound—one stroke—*Pancratius dies no more!*
               Eleanor C. Donnelly.

CATHERINE AND GRIFFITH.

Gri. How does your grace?

Cath. O Griffith, sick to death:
My legs like loaded branches bow to th' earth,
Willing to leave their burden: reach a chair.—
So—now methinks I feel a little ease.
Didst thou not tell me, Griffith, as thou led'st me,
That the great child of honor, Cardinal Wolsey,
Was dead.

Grif. Yes, madam; but I think your grace,
Out of the pain you suffer'd, gave no ear to 't.

Cath. Pr'ythee, good Griffith, tell me how he
died?
If well, he stept before me, happily,
For my example.

Grif. Well, the voice goes, madam,
For after the stout Earl of Northumberland
Arrested him at York, and brought him forward
(As a man sorely tainted) to his answer,
He fell sick suddenly, and grew so ill
He could not sit his mule.

Cath. Alas, poor man!

Grif. At last, with easy roads he came to Lei-
cester,
Lodg'd in the abbey; where the reverend abbot,
With all his convent, honorably receiv'd him;
To whom he gave these words: "O Father Abbot,
An old man broken with the storms of state,
Is come to lay his weary bones among ye;
Give him a little earth for charity!"
So went to bed: where eagerly his sickness

Pursued him still, and three nights after this,
About the hour of eight (which he himself
Foretold should be his last), full of repentance,
Continual meditations, tears and sorrows,
He gave his honors to the world again,
His blessed part to heaven, and slept in peace.

 Cath. So may he rest, his faults lie buried with
  him!
Yet thus far, Griffith, give me leave to speak him,
And yet with charity; he was a man
Of an unbounded stomach, ever ranking
Himself with princes; one that by suggestion
Tithed all the kingdom; simony was fair play:
His own opinion was his law. I' th' presence
He would say untruths, and be ever double
Both in his words and meaning. He was never,
But where he meant to ruin, pitiful.
His promises were, as he then was, mighty;
But his performance, as he now is, nothing.

 Grif. Noble madam,
Men's evil manners live in brass, their virtues
We write in water. May it please your highness
To hear me speak his good now?

 Cath. Yes, good Griffith,
I were malicious else.

 Grif. This Cardinal,
Though from an humble stock, undoubtedly
Was fashion'd to much honor from his cradle:
He was a scholar, and a ripe and good one;
Exceeding wise, fair spoken, and persuading;
Lofty, and sour to them that lov'd him not,

But to those men that sought him sweet as sum-
    mer.
And though he was unsatisfied in getting,
(Which was a sin) yet in bestowing, madam,
He was most princely; ever witness for him,
Those twins of learning that he rais'd in you,
Ipswich and Oxford! one of which fell with him,
Unwilling to outlive the good he did it:
The other, though unfinished, yet so famous,
So excellent in art, and still so rising,
That Christendom shall ever speak his virtue;
His overthrow heaped happiness upon him;
For then, and not till then, he felt himself,
And found the blessedness of being little:
And to add greater honors to his age
Than man could give him, he died, fearing God.
   Cath.   After my death I wish no other herald,
No other speaker, of my living actions,
To keep mine honor from corruption,
But such an honest chronicler as Griffith.
Whom I most hated living, thou hast made me,
With thy religious truth and modesty,
Now in his ashes honor. Peace be with him!
                       Shakespeare.

GUALBERTO'S VICTORY.

A MOUNTAIN pass so narrow that a man
Riding that way to Florence, stooping, can
Touch with his hand the rock, on either side,
And pluck the flowers that in the crannies hide.
Here, on Good Friday, centuries ago,
Mounted and armed, John Gualbert met his foe,

Mounted and armed as well, but riding down
To the fair city from the woodland brown,
This way and that swinging his jewelled whip,
A gay old love-song on his careless lip,
And on his charger's neck the reins loose thrown.

An accidental meeting; but the sun
Burned on their brows, as if it had been one
Of deep design, so deadly was the look
Of mutual hate their olive faces took,
As (knightly courtesy forgot in wrath)
Neither would yield his enemy the path.
"Back!" cried Gualberto. "Never!" yelled his
    foe;
And on the instant, sword in hand, they throw
Them from their saddles, nothing loath,
And fall to fighting, with a smothered oath.
A pair of shapely, stalwart cavaliers,
Well-matched in stature, weapons, weight, and
    years,
Theirs was a long, fierce struggle on the grass,
Thrusting-and parrying up and down the pass;
Swaying from left to right, in combat clenched.
Till all the housings of their steeds were drenched
With brutal gore, and ugly blood-drops oozed
Upon the rocks, from head and hands contused.
But at the close, when Gualbert stopped to rest,
His heel was planted on his foeman's breast;
And looking up, the fallen courtier sees,
As in a dream, gray rocks and waving trees
Before his glazing vision faintly float,
While Gualbert's sabre glitters at his throat.

"Now die, base wretch!" the victor fiercely cries,
His heart of hate outflashing from his eyes:
" Never again, by the all-righteous Lord!
Shalt thou with life escape this trusty sword,—
Revenge is sweet!" And upward glanced the steel,
But ere it fell,—dear Lord! a silvery peal
Of voices chanting in the town below,
Grave, ghostly voices chanting far below,
Rose, like a fountain's spray from spires of
    snow,
And chimed and chimed to die in echoes slow.

In the sweet silence following the sound,
Gualberto and the man upon the ground
Glared at each other with bewildered eyes
(The glare of hunted deer on leashèd hound);
And then the vanquished, struggling to arise,
Made one last effort, while his face grew dark
With pleading agony: " Gualberto! hark!
The chants—the hour—thou know'st the olden
    fashion,—
The monks below intone our Lord's dear Pas-
    sion.
Oh! by this cross!"—and here he caught the hilt
Of Gualbert's sword,—" and by the Blood once
    spilt
Upon it for us both long years ago,
Forgive—forget—and spare a fallen foe!"

The face that bent above grew white and set
(Christ or the demon?—in the balance hung):
The lips were drawn,—the brow bedewed with
    sweat,—

But on the grass the harmless sword was flung:
And stooping down, the hero, generous, wrung
The outstretched hand.   Then, lest he lose con-
    trol
Of the but half-tamed passions of his soul,
Fled up the pathway, tearing casque and coat
To ease the tempest throbbing at his throat;
Fled up the crags, as if a fiend pursued,
And paused not till he reached a chapel rude.

There, in the cool dim stillness, on his knees,
Trembling, he flings himself, and, startled, sees
Set in the rock a crucifix antique,
From which the wounded Christ bends down to
    speak.
  " *Thou hast done well, Gualberto.   For My sake*
Thou didst forgive thine enemy ; now take
My gracious pardon for thy times of sin,
And from this day a better life begin."

White flashed the angels' wings about his head,
Rare, subtile perfumes through the place were
    shed;
And golden harps and sweetest voices poured
Their glorious hosannas to the Lord,
Who in that hour, and in that chapel quaint,
Changed by His power, by His dear love's con-
    straint,
Gualbert the sinner into John the saint.

                  Eleanor C. Donnelly.

QUEEN ARCHIDAMIA.[1]

"Pyrrhus next advanced against the city. It was resolved to send the women into Crete, but they remonstrated against it; and Queen Archidamia, being appointed to speak for the rest, went into the council hall with a sword in her hand, and said that they did their wives great wrong if they thought them so faint-hearted as to live after Sparta was destroyed."

THE chiefs were met in the council hall;
   Their words were sad and few;
They were ready to fight, and ready to fall
   As the sons of heroes do.

And moored in the harbor Gythium lay
   The last of the Spartan fleet,
That should bear the Spartan women away
   To the sunny shores of Crete.

Their hearts went back to the days of old;
   They thought of the world-wide shock,
When the Persian host like an ocean rolled
   To the foot of the Grecian rock.

And they turned their faces, eager and pale,
   To the rising roar in the street,
As if the clank of the Spartan mail
   Were the tramp of the conquerors' feet.

It was Archidamia, the Spartan queen,—
   Brave as her father's steel;
She stood like the silence that comes between
   The flash and the thunder's peal.

[1] This recitation must be given with power and dignity.

She looked in the eyes of the startled crowd;
   Calmly she gazed around;
Her voice was neither low nor loud,
   But it rang like her sword on the ground.

"Spartans," she said, and her woman's face
   Flushed out both pride and shame,
"I ask, by the memory of your race,
   Are ye worthy of your name?

"Ye have bidden us seek new hearths and graves,
   Beyond the reach of the foe;
And now, by the dash of the blue sea waves,
   We swear that we will not go!

"Is the name of Pyrrhus to blanch your cheeks?
   Shall he burn, and kill, and destroy?
Are ye not sons of the deathless Greeks
   Who fired the gates of Troy?

"What though his feet have scathless stood
   In the rush of the Punic foam?
Though his sword be red to its hilt with the
      blood
   That has beat at the heart of Rome?

"Brothers and sons! we have reared you men;
   Our walls are the ocean swell;
Our winds blow keen down the rocky glen
   Where the staunch Three Hundred fell.

" Our hearts are drenched in the wild sea flow,
 In the light of the hills and the sky ;
And the Spartan women, if need be so,
 Will teach the men to die.

" We are brave men's mothers, and brave men's
  wives ;
 We are ready to do and dare ;
We are ready to man your walls with our lives,
 And string your bows with our hair.

" Let the young and brave lie down to-night,
 And dream of the brave old dead,
Their broad shields bright for to-morrow's fight,
 Their swords beneath their head.

" Our breasts are better than bolts and bars ;
 We neither wail nor weep ;
We will light our torches at the stars,
 And work while our warriors sleep.

" We hold not the iron in our own blood
 Viler than strangers' gold ;
The memory of our motherhood
 Is not to be bought and sold.

" Shame to the traitor heart that springs
 To the faint soft arms of peace,
If the Roman eagle shook his wings
 At the very gates of Greece.

"Ask not the mothers who gave you birth
  To bid you turn and flee ;
When Sparta is banished from the earth
  Her women can die and be free."

PROLOGUE TO CATO.

To wake the soul by tender strokes of art,
To raise the genius, and to mend the heart,
To make mankind in conscious virtue bold,
Live o'er each scene, and be what they behold ;
For this the tragic muse first trod the stage,
Commanding tears to stream through every age :
Tyrants no more their savage nature kept,
And foes to virtue wondered how they wept.
Our author shuns by vulgar springs to move
The hero's glory and the virgin's love ;
In pitying love, we but our weakness show,
And wild ambition well deserves its woe.
Here tears shall flow from a more generous
    cause,
Such tears as patriots shed for dying laws :
He bids your breasts with ancient ardor rise,
And calls forth Roman drops from British eyes.
Virtue confessed in human shape he draws,
What Plato taught, and godlike Cato was :
No common object to your sight displays,
But what with pleasure Heav'n itself surveys :
A brave man struggling in the storms of fate,
And greatly falling with a falling state !
While Cato gives his little senate laws,
What bosom beats not in his country's cause ?

Who sees him act, but envies ev'ry deed?
Who hears him groan, and does not wish to
    bleed?
Ev'n when proud Cæsar, midst triumphal cars,
The spoils of nations, and the pomp of wars,
Ignobly vain, and impotently great,
Showed Rome her Cato's figure drawn in state;
As her dead father's rev'rend image past,
The pomp was darkened, and the day o'ercast,
The triumph ceased—tears gushed from ev'ry
    eye,
The world's great victor passed unheeded by;
Her last good man dejected Rome adored,
And honored Cæsar's less than Cato's sword.

                                Pope.

CATO'S SENATE.

    Cato. Fathers, we once again are met in
    council.
Cæsar's approach has summoned us together,
And Rome attends her fate from our resolves;
How shall we treat this bold aspiring man?
Success stills follows him, and backs his crimes:
Pharsalia gave him Rome. Egypt has since
Received his yoke, and the whole Nile is Cæsar's.
Why should I mention Juba's overthrow,
And Scipio's death? Numidia's burning sands
Still smoke with blood. 'Tis time we should de-
    cree
What course to take. Our foe advances on us,
And envies us even Libya's sultry deserts.

Fathers, pronounce your thoughts : are they still
    fixed
To hold it out and fight it to the last?
Or are your hearts subdued at length, and
    wrought
By time and ill success to a submission?
Sempronius, speak.
    Sempronius. My voice is still for war.
Gods! can a Roman senate long debate
Which of the two to choose, slav'ry or death?
No, let us rise at once, gird on our swords,
And at the head of our remaining troops
Attack the foe, break through the thick array
Of his thronged legions, and charge home upon
    him.
Perhaps some arm, more lucky than the rest,
May reach his heart, and free the world from
    bondage.
Rise, fathers, rise! 'tis Rome demands your help;
Rise, and revenge her slaughtered citizens,
Or share their fate! The corpse of half her sen-
    ate
Manure the fields of Thessaly, while we
Sit here delib'rating in cold debates
If we should sacrifice our lives to honor,
Or wear them out in servitude and chains.
Rouse up for shame! Our brothers of Pharsalia
Point at their wounds, and cry aloud, To battle!
Great Pompey's shade complains that we are
    slow,
And Scipio's ghost walks unrevenged among us!
    Cato. Let not a torrent of impetuous zeal

Transport thee thus beyond the bounds of
    reason :
True fortitude is seen in great exploits
That justice warrants and that wisdom guides :
All else is tow'ring frenzy and distraction.
Are not the lives of those who draw the sword
In Rome's defence entrusted to our care ?
Should we thus lead them to a field of slaughter,
Might not th' impartial world with reason say,
We lavished at our deaths the blood of thousands,
To grace our fall, and make our ruin glorious ?
Lucius, we next would know what's your opinion.
    Lucius. My thoughts, I must confess, are turned
    on peace.
Already have our quarrels filled the world
With widows and with orphans ; Scythia mourns
Our guilty wars, and earth's remotest regions
Lie half unpeopled by the feuds of Rome :
'Tis time to sheath the sword, and spare man-
    kind.
It is not Cæsar, but the gods, my fathers,
The gods declare against us, and repel
Our vain attempts. To urge the foe to battle,
(Prompted by blind revenge, and wild despair)
Were to refuse th' awards of Providence,
And not to rest in Heaven's determination.
Already have we shown our love to Rome,
Now let us show submission to the gods.
We took up arms, not to revenge ourselves,
But free the commonwealth ; when this end fails,
Arms have no further use : our country's cause,

That drew our swords, now wrests 'em from our
    hands,
And bids us not delight in Roman blood,
Unprofitably shed ; what men could do
Is done already ; heaven and earth will witness,
If Rome must fall, that we are innocent.
    Sem. This smooth discourse and mild behav-
      ior oft
Conceal a traitor. Something whispers me
All is not right. Cato, beware of Lucius.
    Cato. Let us appear not rash nor diffident ;
Immod'rate valor swells into a fault ;
And fear admitted into public councils
Betrays like treason. Let us shun 'em both.
Fathers, I cannot see that our affairs
Are grown thus desp'rate ; we have bulwarks
    round us :
Within our walls are troops inured to toil
In Afric's heats, and seasoned to the sun ;
Numidia's spacious kingdom lies behind us,
Ready to rise at its young prince's call.
While there is hope, do not distrust the gods :
But wait at least till Cæsar's near approach
Force us to yield. 'Twill never be too late
To sue for chains, and own a conqueror.
Why should Rome fall a moment ere her time ?
No, let us draw our term of freedom out
In its full length, and spin it to the last,
So shall we gain still one day's liberty :
And let me perish, but in Cato's judgment,
A day, an hour of virtuous liberty,
Is worth a whole eternity in bondage.

Enter MARCUS.

Marc. Fathers, this moment, as I watched the
    gate,
Lodged on my post, a herald is arrived
From Cæsar's camp, and with him comes old
    Decius,
The Roman knight: he carries in his looks
Impatience, and demands to speak with Cato.
    Cato. By your permission, fathers, bid him
    enter.
Decius was once my friend, but other prospects
Have loosed those ties, and bound him fast to
    Cæsar.
His message may determine our resolves.

Enter DECIUS.

    Dec. Cæsar sends health to Cato—
    Cato. Could he send it
To Cato's slaughtered friends it would be wel-
    come.
Are not your orders to address the senate?
    Dec. My business is with Cato; Cæsar sees
The straits to which you're driven; and, as he
    knows
Cato's high worth, is anxious for your life.
    Cato. My life is grafted on the fate of Rome.
Would he save Cato, bid him spare his country.
Tell your dictator this: and tell him, Cato
Disdains a life which he has power to offer.
    Dec. Rome and her senators submit to Cæsar;
Her gen'rals and her consuls are no more,

Who checked his conquest, and denied his tri-
    umphs.
Why will not Cato be this Cæsar's friend?
    Cato. Those very reasons thou hast urged for-
    bid it.
    Dec. Cato, I've orders to expostulate,
And reason with you, as from friend to friend ;
Think on the storm that gathers o'er your head,
And threatens ev'ry hour to burst upon it ;
Still may you stand high in your country's honors,
Do but comply, and make your peace with Cæsar.
Rome will rejoice, and cast its eyes on Cato,
As on the second of mankind.
    Cato. No more :
I must not think of life on such conditions.
    Dec. Cæsar is well acquainted with your virtues,
And therefore sets this value on your life :
Let him but know the price of Cato's friendship,
And name your terms.
    Cato. Bid him disband his legions,
Restore the commonwealth to liberty,
Submit his actions to the public censure,
And stand the judgment of a Roman senate.
Bid him do this, and Cato is his friend.
    Dec. Cato, the world talks loudly of your wis-
    dom—
    Cato. Nay more, tho' Cato's voice was ne'er
    employed
To clear the guilty, and to varnish crimes,
Myself will mount the Rostrum in his favor,
And strive to gain his pardon from the people.
    Dec. A style like this becomes a conqueror.

Cato. Decius, a style like this becomes a Roman.

Dec. What is a Roman that is Cæsar's foe?

Cato. Greater than Cæsar: he's a friend to virtue.

Dec. Consider, Cato, you're in Utica,
And at the head of your own little senate;
You don't now thunder in the Capitol,
With all the mouths of Rome to second you.

Cato. Let him consider that, who drives us hither.
'Tis Cæsar's sword has made Rome's senate little,
And thinned its ranks. Alas! thy dazzled eye
Beholds this man in a false glaring light,
Which conquest and success have thrown upon him;
Did'st thou but view him right, thou'dst see him black
With murder, treason, sacrilege, and crimes
That strike my soul with horror but to name 'em.
I know thou look'st on me as on a wretch
Beset with ills, and covered with misfortunes;
But, by the gods I swear, millions of worlds
Should never buy me to be like that Cæsar.

Dec. Does Cato send this answer back to Cæsar,
For all his gen'rous cares, and proffered friendship?

Cato. His cares for me are insolent and vain;
Presumptuous man! the gods take care of Cato.
Would Cæsar show the greatness of his soul,

Bid him employ his care for these my friends,
And make good use of his ill-gotten power,
By shelt'ring men much better than himself.
   Dec. Your high unconquered heart makes you
       forget
You are a man.   You rush on your destruction.
But I have done.   When I relate hereafter
The tale of this unhappy embassy,
All Rome will be in tears.

Addison.

THE BATTLE OF WATERLOO.

THERE was a sound of revelry by night,
And Belgium's capital had gathered then
Her beauty and her chivalry, and bright
The lamps shone o'er fair women and brave
    men ;
A thousand hearts beat happily ; and when
Music rose with its voluptuous swell,
Soft eyes looked love to eyes which spake again,
And all went merry as a marriage-bell ;
But hush ! hark ! a deep sound strikes like a
    rising knell !

Did ye not hear it ?—No, 'twas but the wind,
Or the car rattling o'er the stony street ;
On with the dance ! let joy be unconfined ;
No sleep till morn, when youth and pleasure
    meet
To chase the glowing hours with flying feet—
But hark ! that heavy sound breaks in once
    more,

As if the clouds its echo would repeat ;
And nearer, clearer, deadlier than before !
Arm ! arm ! it is—it is—the cannon's opening
   roar !

Ah ! then and there was hurrying to and fro,
And gathering tears, and tremblings of distress,
And cheeks all pale, which but an hour ago
Blushed at the praise of their own loveliness.
And there were sudden partings, such as press
The life from out young hearts, and choking
   sighs
Which ne'er might be repeated ; who could
   guess
If evermore should meet those mutual eyes,
Since upon night so sweet such awful morn
   could rise !

And there was mounting in hot haste the steed ;
The mustering squadron, and the clattering car
Went pouring forward with impetuous speed,
And swiftly forming in the ranks of war ;
And the deep thunder peal on peal afar :
And near, the beat of the alarming drum
Roused up the soldier ere the morning star ;
While thronged the citizens with terror dumb,
Or whispering, with white lips—" The foe ! they
   come ! they come !"

Last noon beheld them full of lusty life,
Last eve in beauty's circle proudly gay,
The midnight brought the signal sound of strife,

The morn the marshalling in arms—the day
Battle's magnificently stern array !
The thunder-clouds close o'er it, which, when
rent,
The earth is covered thick with other clay,
Which her own clay shall cover, heaped and
pent,
Rider and horse—friend, foe—in one red burial
blent !

Byron.

THE BRIDAL OF MALAHIDE.

THE joy-bells are ringing in gay Malahide,
The fresh wind is singing along the sea-side ;
The maids are assembling with garlands of flow-
ers,
And the harp-strings are trembling in all the glad
bowers.

Swell, swell the gay measure ! roll trumpet and
drum !
'Mid greetings of pleasure in splendor they
come !
The chancel is ready, the portal stands wide,
For the lord and the lady, the bridegroom and
bride.

Before the high altar young Maud stands ar-
rayed !
With accents that falter her promise is made—
From father and mother forever to part,
For him and no other to treasure her heart.

The words are repeated, the bridal is done,
The rite is completed—the two, they are one ;
The vow, it is spoken all pure from the heart,
That must not be broken till life shall depart.

Hark ! 'mid the gay clangor that compassed their
    car,
Loud accents in anger come mingling afar !
The foe's on the border ! his weapons resound
Where the lines in disorder unguarded are found !

As wakes the good shepherd, the watchful and
    bold,
When the ounce or the leopard is seen in the
    fold,
So rises already the chief in his mail,
While the new-married lady looks fainting and
    pale.

"Son, husband, and brother, arise to the strife,
For sister and mother, for children and wife !
O'er hill and o'er hollow, o'er mountain and
    plain,
Up, true men, and follow ! let dastards remain !"

Farrah ! to the battle ! They form into line—
The shields, how they rattle ! the spears, how
    they shine !
Soon, soon shall the foeman his treachery rue—
On, burgher and yeoman ! to die or to do !

The eve is declining in lone Malahide :
The maidens are twining gay wreaths for the
    bride ;

She marks them unheeding—her heart is afar,
Where the clansmen are bleeding for her in the
war.

Hark! loud from the mountain—'tis victory's
cry !
O'er woodland and fountain it rings to the sky !
The foe has retreated ! he flees to the shore;
The spoiler's defeated—the combat is o'er !

With foreheads unruffled the conquerors come—
But why have they muffled the lance and the
drum?
What form do they carry aloft on his shield ?
And where does *he* tarry, the lord of the field ?

Ye saw him at morning, how gallant and gay !
In bridal adorning, the star of the day :
Now, weep for the lover—his triumph is sped,
His hope, it is over !—the chieftain is dead !

But, oh! for the maiden who mourns for that
chief,
With heart overladen, and broken with grief !
She sinks on the meadow :—in one morning-tide,
A wife and a widow, a maid and a bride !

Ye maidens attending, forbear to condole !
Your comfort is rending the depths of her soul.
True—true, 'twas a story for *ages* of pride ;
He died in his glory—but, oh, he *has* died !

Gerald Griffin (*Altered*).

THE MOURNERS.

KING DEATH sped forth in his dreaded power
To make the most of his tyrant hour;
And the first he took was a white-robed girl,
With the orange bloom twined in each glossy
  curl.
Her fond betrothed hung over the bier,
Bathing her shroud with the gushing tear:
He madly raved, he shrieked his pain,
With frantic speech and burning brain.
"There's no joy," cried he, "now my dearest is
  gone,
Take, take me, Death; for I cannot live on!"

The valued friend, too, was snatched away,
Bound to another from childhood's day;
And the friend that was left exclaimed in despair,
"Oh! he sleeps in the grave—let me follow him
  there!"

A mother was taken, whose constant love
Had nestled her child like a fair young dove;
And the heart of that child to the mother had
  grown
Like the ivy to oak, or moss to the stone;
Nor loud nor wild was the burst of woe,
But the tide of anguish ran strong below;
And the reft one turned from all that was light,
From the flowers of day and the stars of night;
Breathing where none might hear or see—
"Where thou art, my mother, thy child would
  be."

Death smiled as he heard each earnest word :
" Nay, nay," said he, " be this work deferred ;
I'll see you again in a fleeting year,
And, if grief and devotion live on sincere,
I promise then ye shall share the rest
Of the beings now plucked from your doting
      breast ;
Then, if ye crave still the coffin and pall
As ye do this moment, my spear shall fall."
And Death fled till time on his rapid wing
Again brought back the skeleton king.

But the lover was ardently wooing again,
Kneeling in serfdom, and proud of his chain ;
He had found an idol again to adore,
Rarer than that he had worshipped before :
His step was gay, his laugh was loud,
As he led the way for the bridal crowd ;
And his eyes still kept their joyous ray,
Though he went by the grave where his first love
      lay.
" Ha ! ha !" shouted Death, " 'tis passing clear
That I am a guest not wanted here !"

The friend again was quaffing the bowl,
Warmly pledging his faith and soul ;
His bosom cherished with glowing pride
A stranger form that sat by his side ;
His hand the hand of that stranger pressed ;
He praised his song, he echoed his jest ;
And the mirth and wit of that new-found mate
Made a blank of the name so prized of late.

"See! see!" cried Death, as he hurried past,
" How bravely the bonds of friendship last !"

But the orphan child! Oh, where was she?
With clasping hands and bended knee,
All alone on the churchyard sod,
Mingling the names of mother and God.
Her dark and sunken eye was hid,
Fast weeping beneath the sunken lid ;
Her sighs were heavy, her forehead was chill,
Betraying the wound was unhealed still ;
And her smothered prayer was yet heard to crave
A speedy home in the self-same grave.

Hers was the love all holy and strong;
Hers was the sorrow fervent and long;
Hers was the spirit whose light was shed
As an incense fire above the dead.
Death lingered there, and paused awhile ;
But she beckoned him on with a welcoming
    smile.
" There's a solace," cried she, " for all others to
    find,
But a mother leaves no equal behind."
And the kindest blow Death ever gave
Laid the mourning child in the mother's grave.
                                        E. Cook.

AN ORDER FOR A PICTURE.

O GOOD painter, tell me true,
  Has your hand the cunning to draw
  Shapes of things that you never saw ?
Ay ? Well, here is an order for you.

Woods and cornfields, a little brown,—
  The picture must not be over-bright,
  Yet all in the golden and gracious light
Of a cloud, when the summer sun is down.
    Alway and alway, night and morn,
  Woods upon woods, with fields of corn
    Lying between them, not quite sere,
And not in the full, thick, leafy bloom,
When the wind can hardly find breathing-room
    Under their tassels,—cattle near,
Biting shorter the short green grass,
And a hedge of sumach and sassafras,
With bluebirds twittering all around,—
(Ah, good painter, you can't paint sound !)
    These, and the house where I was born,
Low and little, and black and old,
With children, many as it can hold,
All at the windows, open wide,—
Heads and shoulders clear outside,
And fair young faces all ablush :
    Perhaps you may have seen, some day,
    Roses crowding the self-same way,
Out of a wilding, wayside bush.

    Listen closer. When you have done
    With woods and cornfields and grazing herds,
  A lady, the loveliest ever the sun
Looked down upon, you must paint for me ;
Oh, if I only could make you see
    The clear blue eyes, the tender smile,
The sovereign sweetness, the gentle grace,
The woman's soul, and the angel's face

That are beaming on me all the while,
I need not speak these foolish words :
Yet one word tells you all I would say,—
She is my mother : you will agree
That all the rest may be thrown away.

Two little urchins at her knee
You must paint, sir ; one like me,
    The other with a clearer brow,
    And the light of his adventurous eyes
    Flashing with boldest enterprise :
At ten years old he went to sea,—
    God knoweth if he be living now ;
    He sailed in the good ship " Commodore,"—
Nobody ever crossed her track
To bring us news, and she never came back.
    Ah, 'tis twenty long years and more
Since that old ship went out of the bay
    With my great-hearted brother on her deck :
    I watched him till he shrank to a speck,
And his face was toward me all the way.
Bright his hair was, a golden brown,
    The time we stood at our mother's knee :
That beauteous head, if it did go down,
    Carried sunshine into the sea !

Out in the fields one summer night
    We were together, half afraid
    Of the corn-leaves' rustling, and of the shade
    Of the high hills, stretching so still and far,—
Loitering till after the low little light

Of the candle shone through the open door,
   Afraid to go home, sir ; for one of us bore
A nest full of speckled and thin-shelled eggs ;
The other, a bird, held fast by the legs,
Not so big as a straw of wheat :
The berries we gave her she wouldn't eat,
But cried and cried, till we held her bill,
So slim and shining, to keep her still.

At last we stood at our mother's knee.
   Do you think, sir, if you try,
   You can paint the look of a lie ?
   If you can, pray have the grace
   To put it solely in the face
Of the urchin that is likest me :
     I think 'twas solely mine, indeed :
    But that's no matter,—paint it so ;
     The eyes of our mother—(take good heed)—
Looking not on the nest full of eggs,
Nor the fluttering bird, held so fast by the legs,
But straight through our faces down to our lies,
And oh, with such injured, reproachful surprise !
   I felt my heart bleed where that glance went,
      as though
   A sharp blade struck through it.
                  You, sir, know
That you on the canvas are to repeat
Things that are fairest, things most sweet,—
Woods and cornfields and mulberry tree,—
The mother,—the lads, with their bird, at her
     knee :

But, oh, that look of reproachful woe !
High as the heavens your name I'll shout,
    If you paint me the picture, and leave that out.
                                    Alice Cary.

THE ROSARY OF MY TEARS.

SOME reckon their ages by years,
    Some measure their life by art—
But some tell their days by the flow of their
    tears,
And their life by the moans of their heart.

The dials of earth may show
    The length, not the depth of years,
Few or many they come, few or many they go—
But our time is best measured by fears.

Ah ! not by the silver gray
    That creeps through the sunny hair,
And not by the scenes that we pass on our way—
And not by the furrows the finger of care

On the forehead and face have made—
    Not so do we count our years ;
Not by the sun of the earth, but the shade
Of our souls, and the fall of our tears.

For the young are oft-times old,
    Though their brow be bright and fair ;
While their blood beats warm their heart lies
    cold—
O'er them the spring-time, but winter is there.

And the old are oft-times young
When their hair is thin and white,
And they sing in age as in youth they sung,
And they laugh, for their cross was light.

But bead by bead I tell
The rosary of my years;
From a cross to a cross they lead—'tis well!
And they're blessed with a blessing of tears.

Better a day of strife
Than a century of sleep;
Give me instead of a long stream of life
The tempest and tears of the deep.

A thousand joys may foam
On the billows of all the years;
But never the foam brings the brave bark home:
It reaches the haven through tears.

Father Ryan.

LABOR.

THERE is a perennial nobleness, and even
sacredness, in work. Were a man ever so be-
nighted, or forgetful of his high calling, there is
always hope in him who actually and earnestly
works; in idleness alone is there perpetual despair. Consider how, even in the meanest sorts
of labor, the whole soul of a man is composed
into real harmony. He bends himself with free
valor against his task; and doubt, desire, sorrow,
remorse, indignation, despair itself, shrink murmuring far off into their caves. The glow of

labor in him is a purifying fire, wherein all poison is burnt up; and of smoke itself there is made a bright and blessed flame.

Blessed is he who has found his work; let him ask no other blessedness; he has a life purpose. Labor is life. From the heart of the worker rises the celestial force, breathed into him by Almighty God, awakening him to all nobleness, to all knowledge. Hast thou valued patience, courage, openness to light, or readiness to own thy mistakes? In wrestling with the dim, brute powers of Fact, thou wilt continually learn. For every noble work, the possibilities are diffused through immensity—undiscoverable, except to Faith.

Man, son of heaven! is there not in thine inmost heart a spirit of active method, giving thee no rest till thou unfold it? Complain not. Look up, wearied brother. See thy fellow-workmen surviving through eternity—the sacred band of immortals!

Thomas Carlyle.

THE RUSTIC BRIDAL;

OR, THE BLIND GIRL OF CASTEL CUILLE.

AT the foot of the mountain height
Where is perched the Castel Cuille,
When the apple, the plum, and the almond tree
In the plain below were growing white,
This is the song one might perceive
On a Wednesday morn of St. Joseph's Eve:

"The roads should blossom, the roads should
    bloom,
So fair a bride shall leave her home ;
Should blossom and bloom with garlands gay,
So fair a bride shall pass to-day."
This old Te Deum, rustic rites attending,
Seemed from the clouds descending,
When lo! a merry company
Of rosy village girls, clean as the eye,
Each one with her attendant swain,
Came to the cliff, all singing the same strain ;
Resembling there, so near unto the sky,
Rejoicing angels, that kind Heaven has sent
For their delight and our encouragement.
        Together blending,
        And soon descending
        The narrow sweep
        Of the hillside steep,
        They wind aslant
        Toward Saint Amant,
        Through leafy alleys
        Of verdurous valleys
        With merry sallies
        Singing their chant :

"The roads should blossom, the roads should
    bloom,
So fair a bride shall leave her home,
Should blossom and bloom with garlands gay,
So fair a bride shall pass to-day."

It is Baptiste, and his affianced maiden,
With garlands for the bridal laden.

> Gayly frolicking,
> Wildly rollicking !
> > Kissing,
> > Caressing,
> With fingers pressing,
> Till in the veriest
> Madness of mirth, as they dance,
> They retreat and advance,
> Trying whose laugh shall be loudest and
> > merriest.
> Meanwhile Baptiste stands sighing, with silent
> > tongue ;
> And yet the bride is fair and young !

> Now you must know one year ago,
> That Margaret, the young and tender,
> Was the village pride and splendor ;
> But, alas ! the summer's blight,
> That dread disease that none can stay,
> The pestilence that walks by night,
> Took the young bride's sight away.
> Bereft of joy, ere long the lover fled ;
> Returned but three days ago,
> The golden chain they round him throw,
> He is enticed, and onward led
> To marry Angela, and yet
> Is thinking ever of Margaret.

> But here comes crippled Jane, the village seer,
> She wears a countenance severe,
> And saith, " When Angela weddeth this false
> > bridegroom,
> She diggeth for Margaret a tomb."

Beautiful as some fair angel yet,
Thus lamented Margaret:
" He has arrived; arrived at last,
Yet Jane has named him not these three days
     past;
But some one comes!   Though blind, my heart
     can see!
And that deceives me not! 'tis he! 'tis he!"
With outstretched arms, but sightless eyes,
She rises; 'tis only Paul, her brother, who thus
     cries:
" Angela, the bride, has passed.
Tell me, my sister, why were we not asked?"
" Angela married! and not send
To tell her secret unto me!
Oh! speak! who may the bridegroom be?"
" My sister, 'tis Baptiste, thy friend!"
A cry the blind girl gave, but nothing said;
A milky whiteness upon her cheek is spread.
" Hark! the joyous airs are ringing!
Sister, dost thou hear them singing?
How merrily they laugh and jest!
Would we were bidden with the rest!
I would don my hose of homespun gray,
And my doublet of linen, striped and gay.
Perhaps they will come; for they do not wed
Till to-morrow at seven, it is said!"
" Paul, be not sad! 'tis a holiday;
To-morrow put on thy doublet gay;
But leave me now for awhile alone."

Away with a hop and a jump went Paul,
And as he whistled along the hall,

Entered Jane, the crippled crone.
"I'm faint! What dreadful heat!
My little friend! what ails thee, sweet?"
"Nothing! I heard them singing home the bride;
And as I listened to the song,
I thought my turn would come ere long:
Thou knowest it is at Whitsuntide."
Jane, shuddering, her hand doth press;
"Thy love I cannot all approve;
We must not trust too much to happiness:
Go, pray to God that thou may'st love him less."
"The more I pray the more I love!
It is no sin, for God is on my side!"
It was enough, and Jane no more replied,
But when departing at the evening's close,
She murmured, "She may be saved, she nothing
     knows!"
Now rings the bell, nine times reverberating.
And the white daybreak, stealing up the sky,
Sees in two cottages two maidens waiting,
       How differently!
       The one fantastic, light as air,
         'Mid kisses ringing
         And joyous singing
       Forgets to say her morning prayer!
The other, with cold drops upon her brow,
Joins her two hands and kneels upon the floor,
And whispers, as her brother opes the door,
"O God! forgive me now!"
And then the orphan, young and blind,
Conducted by her brother's hand,
Toward the church, through paths unscanned,

With tranquil air her way doth wind.
"Paul," said Margaret, "where are we? we as-
    cend!"
" Yes, we are at our journey's end!
Come in! The bride will be here soon;
Thou tremblest! O Margaret! art going to
    swoon?"
But no more restrained, no more afraid,
She walks, as for a feast arrayed,
And in the ancient chapel's sombre night
They both are lost to sight.
The guests delay not long,
Soon arrives the village throng.
The wedding-ring is blessed; Baptiste receives it,
Ere on the finger of the bride he leaves it.
He must say one word! 'tis said, and suddenly
    at his side,
" 'Tis he!" a well-known voice hath cried.
And while the wedding guests all hold their breath,
Lo! Margaret, the blind girl, see!
"Baptiste," she said, " since thou hast wished
    my death,
I freely sacrifice myself for thee!"
And calmly in the air a knife suspended.
Doubtless her guardian angel near attended,
For anguish did its work so well,
That ere the fatal stroke descended
Lifeless she fell!

      At eve, instead of bridal verse,
      The De Profundis filled the air;
      Decked with flowers a simple hearse
      To the churchyard forth they bear.

Village girls in robes of snow
Follow, weeping as they go;
Nowhere was a smile that day,
No, ah! no! for each one seemed to say:

"The road should mourn and be veiled in gloom,
So fair a corpse shall leave its home;
Should mourn and should weep, ah! well-away!
So fair a corpse shall pass to-day."

Longfellow.

THE BRAVEST BATTLE THAT EVER WAS FOUGHT.

THE bravest battle that ever was fought,
  Shall I tell you where and when?
On the maps of the world you will find it not;
  'Twas fought by the mothers of men.

Nay, not with cannon, or battle-shot,
  With sword, or nobler pen;
Nay, not with eloquent word or thought,
  From mouths of wonderful men.

But deep in a walled-up woman's heart—
  Of woman that would not yield,
But bravely, silently bore her part—
  Lo! there is that battle-field!

No marshalling troop, no bivouac song;
  No banners to gleam and wave!
But oh! these battles they last so long—
  From babyhood to the grave!

Yet faithful still as a bridge of stars,
    She fights in her walled-up town—
Fights on, and on, in the endless wars,
    Then silent, unseen goes down !

Oh ! ye with banners and battle-shot,
    And soldier to shout and praise,
I tell you the kingliest victories fought
    Are fought in these silent ways !

Oh ! spotless woman in a world of shame,
    With splendid and silent scorn,
Go back to God as white as you came,
    The kingliest warrior born.
                    Joaquin Miller.

THE WIVES OF WEINSBERG.

WHICH way to Weinsberg ? neighbor, say !
    'Tis sure a famous city :
It must have cradled, in its day,
Full many a maid of noble clay,
    And matrons wise and witty ;
And if ever marriage should happen to me,
A Weinsberg dame my wife shall be.

King Conrad once, historians say,
    Fell out with this good city :
So down he came, one luckless day,
Horse, foot, dragoons, in stern array,
    And cannon,—more's the pity !
Around the walls the artillery roared,
And bursting bombs their fury poured.

But naught the little town could scare ;
    Then, red with indignation,
He bade the herald straight repair
Up to the gates, and thunder there
    The following proclamation :
" Rascals ! when I your town do take,
No living thing shall save its neck !"

Now, when the herald's trumpet sent
    These tidings through the city,
To every house a death-knell went;
Such murder-cries the hot air rent
    Might move the stones to pity.
Then bread grew dear, and good advice
Could not be had for any price.

Then "Woe is me !"   "Oh, misery !"
    What shrieks of lamentation !
And " Kyrie Eleison !" cried
The pastors, and the flock replied,
    "Lord, save us from starvation !"
" Oh, woe is me, poor Corydon !
My neck ! my neck ! I'm gone ! I'm gone !"

Yet oft, when counsel, deed, and prayer
    Had all proved unavailing,
When hope hung trembling on a hair,
How oft has woman's wit been there,
    A refuge never-failing !
For woman's wit to conquer fraud
In olden time was famed abroad.

A youthful dame,—praised be her name !
    Last night had seen her plighted,—
Whether in waking hour or dream,
Conceived a rare and novel scheme,
    Which all the town delighted ;
Which you, if you think otherwise,
Have leave to laugh at and despise.

At midnight hour, when culverin
    And gun and bomb were sleeping,
Before the camp, with mournful mien,
The loveliest embassy was seen
    All kneeling low and weeping.
So sweetly, plaintively they prayed,
But no reply save this was made :

" The women have free leave to go,
    Each with her choicest treasure ;
But let the knaves, their husbands, know
That unto them the king will show
    The weight of his displeasure."
With these sad terms the lovely train
    Stole weeping from the camp again.

But when the morning gilt the sky,
    What happened ?　Give attention.
The city gates wide open fly,
And all the wives came trudging by,
    Each bearing—need I mention ?—
Her own dear husband on her back,
All snugly seated in a sack !

Full many a sprig of court, the joke
    Not relishing, protested,
And urged the king ; but Conrad spoke :
" A monarch's word must not be broke !"
    And here the matter rested.
" Bravo !" he cried. " Ha ! ha ! bravo !
Our lady guessed it would be so."

He pardoned all, and gave a ball
    That night at royal quarters.
The fiddles squeaked, the trumpets blew,
And up and down the dancers flew,
    Court sprigs with city daughters.
The major's wife—oh, rarest sight !—
Danced with the shoemaker that night !

Ah, where is Weinsberg, sir, I pray ?
    'Tis sure a famous city :
It must have cradled, in its day,
Full many a maid of noble clay,
    And matrons wise and witty ;
And if ever marriage should happen to me,
A Weinsberg dame my wife shall be.
                  Gottfried August Bürger.

LITTLE JOE.

TOM WISE, a great, big, handsome fellow, with
a heart of the same order, was standing at the
corner talking to a friend. He held a cigar to
his mouth with his left hand, and with his right
had just struck a match against the lamp-post,
when at—or rather under—his elbow a voice exclaimed, cheerily, " Busted ag'in, Mas' Tom !"

Tom threw a glance over his shoulder, and there stood "Little Joe," a small misshapen negro, about fifteen years old, with crutches under his arms, and feet all twisted out of shape, his toes barely touching the ground as he hopped along. He had on an old straw hat with only a hint of brim. There must be some law of cohesive attraction between straw and wool, for Little Joe's cranium was large, while the hat was small and set back much nearer the nape of his neck, than the crown of his head, yet held its place like a natural excrescence or a horrible bore. Joe had met with very few people mean enough to laugh at him; for, though he possessed all the brightness and cheerfulness and pluck of deformed people generally, there was a wistful look about his eyes which his want of height and his position on crutches intensified (indeed, perhaps created), by keeping them upturned while · talking with any one taller than himself; and this was generally the case, for there were no grown people so small as Little Joe. His shirt was torn and his pantaloons ragged, but to gild these faded glories he wore a swallow-tailed coat with brass buttons which some one had given him, whether from a sense of humor or a sentiment of charity let the gods decide.

"Busted ag'in, Mas' Tom!"

"What 'busted' you this time, Joe?" asked Mr. Wise.

"Lumber, Mas' Tom. I was in de lumber

bizness las' week, buyin' ole shingles an' sellin'
'em for kindlin'; but my pardner he maked a
run on de bank,—leastways on my breeches-
pocket,—an' den runned away hisse'f. Ain't you
gwine to sot me up ag'in, Mas' Tom?"

"What business are you going into this
week?"

"'Feckshunerry," replied Joe, taking the quarter Mr. Wise handed him. "Dis'll do to buy de
goods, but 'twon't rent de sto', Mas' Tom."

"What store?" asked Mr. Wise.

"Dat big sto' Hunt an' Manson is jes' moved
outen. Mr. Manson say I may hab it for seb-
ben hunderd dollars ef you'll go my skoorty."

Tom laughed: "Well, Joe, I was thinking I
wouldn't go security for anybody this week.
Don't you think you can do business on a small-
er scale?"

Joe's countenance fell, and he suffered visibly,
but a cheering thought presently struck him,
and he exclaimed, disdainfully: "Anyhow, I ain't
a-keerin' 'bout Hunt an' Manson's ole sto',—der
ole sebben-hunderd-dollar sto'! I can git a
goods-box, and turn it upside down, an' stan' it
up by de Cap'tol groun's, an' more folks'll pass
'long an' buy goobers dan would come in dat ole
sto' all de year. Dey ain't spitin' *me!*"

As Joe limped off to invest his money, his
poor little legs swinging and his swallow-tails
flapping, Tom's friend asked who he was.

"Belonged to us before the war," said Tom.
"Poor little devil! the good Lord and the birds

of the air seem to take care of him. I set him up in business with twenty-five cents every week, and look after him a little in other ways. Sometimes he buys matches and newspapers and sells them again, sometimes he buys ginger-cakes and eats them all; but he is invariably 'busted,' as he calls it, by Saturday night.—Joe! o-oh, Joe!"

Joe looked back, and, with perfect indifference to the fact that he was detaining Mr. Wise, answered that he would " be dar torectly," continuing his negotiations for an empty goods-box lying at the door of a neighboring dry-goods store. " What you want, Mas' Tom?" he asked, on his return.

" Miss Mollie is going to be married week after next, Joe, and you may come up to the house if you like. I was afraid I might forget it."

"Whoop *you*, sir! Thanky, Mas' Tom. I boun' to see Miss Mollie step off de carpit. But, Lord-a-mussy! dem new niggers you all got ain't gwine to lemme in."

" Come to the front door and ask for me. Cut out, now, and don't get busted this week, because I shall need all my money to buy a breastpin to wait on my sister in.—Come, John, let's register."

Joe's glance followed Mr. Wise and his friend till they were out of sight: then he turned, and paused no more till he reached an out-of-the-way grocery-store, in the window of which were displayed samples of fish, and soap, and calico, and kerosene lamps, and dreadful brass jewelry, among which was a frightful breastpin in the shape of a crescent set with red and green glass, and further ornamented by a chain of the most atrocious description conceivable. Before this thing of beauty, which to him had been a joy for weeks, Joe paused and lingered, and smote his little black breast and sighed the sigh of poverty. Then he went in. "What mout be de price o' dat gent's pin in de corner ob de winder?" he inquired.

"I don't see any gent in the corner of the window," said the proprietor of the store.

Joe took the mild pleasantry, and, inquiring, "What mout be de price o' de pin?" was told that it might be anything,—from nothing up,—but it could go for seventy-five cents.

He stood again outside the window, looking sadly and reflecting at the attractive bijou, then seated himself on the curbstone, his crutches resting in the gutter, and thoughtfully smoothed between his finger and thumb the twenty-five cent note Mr. Wise had given him : "Ef I takes dis, an' de one Mas' Tom gwine to gimme nex' week, dat'll be fifty cents, but it won't be seventy-five : so I got to make a quarter on de two. Ef Miss Mollie knowed, I spec' she would wait anoder week to git married, an' den I wouldn't run no resk o' dese ; but I ain't gwine to tell her, cos I know she couldn't help tellin' Mas' Tom, an' I want to s'prise him. Mas' Tom is made me feel good a many a time : I want to make him feel good wunst. He don't nuvvur come dis way, an' ain't seed dat pin, or he would ha' had it 'fore now."

Then Little Joe bestirred himself, and, obtaining the assistance of a friend, took his dry-goods box up to Capitol Square. There he turned it upside down, spread a newspaper over the top, and proceeded to display his wares.

A pyramid of three apples stood in one corner; a small stack of peppermint candy was its *vis-à-vis;* a tiny glass of peanuts graced the third, and was confronted by a lemon that had seen life and was now more sere than yellow. But the crowning glory was the centre-piece,—an unhappy-looking pie, of visage pale and thin physique, yet how beautiful to Joe! He stepped back on his crutches, turning his head from side to side as he surveyed the effect, took up a locust branch he had brought with him to brush away the flies, and, leaning against the iron railing, with calm dignity awaited coming events.

His glance presently fell on the figure of a negro boy, who stood gazing with longing eyes on the delicacies of his table; and it was with a strange feeling of kinship that Little Joe continued to regard the new-comer, for he too had been branded by misfortune. He appeared about Joe's age, and should have been taller, but his legs had been amputated nearly up to the knee, and as he stood on the pitiful stumps, supported by a short cane in one hand, his head was hardly as high as the iron railing. He had none of Joe's brightness, but looked ragged and dirty and hungry, and evidently had no Mas' Tom to help the good Lord and the birds of the air to take care of him. His skin was of a dull ashen hue, and the short wool which clung close to his scalp was sunburnt till it was red and crisp and formed a curious contrast to his black face. One arm was bare, only the ragged remains of a sleeve hanging over the shoulder, and it seemed no great misfortune that his legs had been shortened, for he had hardly pantaloons enough to cover what he had left.

He looked at the pie, and Joe looked at him. Presently the latter inquired, seriously, "Whar yo' legs?"

"Cut off," was the answer.

"How come dey cut off?"

"Feet was fros'-bit. Like ter kill me."

"What yo' name?" asked Joe.

"Kiah."

"What were yo' ole mas' name?"

"Didn't have no ole mas'."

"Was you a natchul free nigger?"

"Dunno what you mean," said Kiah.

"'Fore we was all *sot* free," explained little Joe. "Was you born wid a ole mas' an' a ole mis', or was you born free?—jes' natchully free."

"Free," said Kiah, thus placing himself, as every Southerner knows, under the ban of Joe's contempt. "Umph! my Lor'! Dat pie sholy do smell good!"

"You look hongry," said Joe gravely.

"I is," said Kiah,—"hongry as a dog!"

Negroes are generous creatures, and Joe's mind was fully made up to give Kiah a piece of pie; but before he signified this benevolent intention he rested his crutches under his shoulders and swung his misshapen feet almost in Kiah's face. He leered at him; he grinned at him; he stuck his chin in his face, and made a dart at him with the crown of his head.

He stopped suddenly and grinned ferociously at Kiah. Kiah gazed stolidly back at Joe. Then Joe stepped to the table, took up a rusty old pocket-knife, and cutting out a piece of the pie, handed it to Kiah. Kiah bit off a point of the triangle with his eyes fixed on Joe as if in doubt whether he would be allowed to proceed, but, finding that liberty was not resented, he eagerly devoured the remainder, drew his coat-sleeve across his mouth, and said, "Thanky." And thus their friendship commenced.

[Joe and Kiah soon became the warmest of friends, and on the evening of the wedding the grateful little fellow surprised and touched his benefactor by the present of the breastpin. This Mr. Wise wore through the ceremony, in spite of all remonstrance to the contrary.]

When, however, as the bridal cortège passed through the hall, he saw Joe nudge a fellow-servant with his elbow and point out the pin, he felt repaid, though Miss Annan was holding her head very high indeed.

The next morning little Joe came by the office: " What did de folks say 'bout yo' bres'pin, Mas' Tom ?"

"Say? Why, they did not know what to say, Joe. They could not take their eyes off me. That pin knocked the black out of everything there. The bridegroom couldn't hold a candle to me," said Mr. Wise; and Joe laughed aloud with delight. "Did they give you your supper?"

"Did dat, Mas' Tom; an' I tuk home a snowball an' a orange to Kiah," said little Joe.

Late on the evening of the same day Mr. Wise was about leaving his office, when Little Joe's crutches sounded in the door-way, and Little Joe himself appeared, sobbing bitterly, tears streaming down his face: "Oh, Lordy, Mas' Tom! oh, Lordy!"

"What is the matter, Joe?"

"Oh, Lordy, Mas' Tom! Kiah's done dead!"

"Kiah! Is it possible? What was the matter?" asked Mr. Wise.

"Oh, Lordy! oh, Lordy!" sobbed Little Joe. "Me an' him went down to de creek, an' was playin' babtizin' an' I'd done babtized Kiah, an' —oh, Lordy! Lordy!—an' Kiah was jes' gwine to babtize me, an' he stepped out too fur, an' his legs was so short he lost his holt on me an' drownded; an' I couldn't ketch him, 'cos I couldn't stan' up widout nothin' to hold on to. Oh, Lordy! I wish I nuvvur had ha' heerd o' babtizin'! I couldn't git him out, an' I jes' kep' on a-hollerin', but nobody didn't come till Kiah was done drownded."

"I am sorry for you, Joe; I wish I had been

there. But, as far as Kiah is concerned, he is better off than he was before," said Mr. Wise.

"No, he ain't, Mas' Tom," said Joe, stoutly; "leas'ways, Kiah didn't think so hisse'f, 'cos ef he had a-wanted to die he could ha' done it long an' merry ago. I don't b'leeve in no sech fool-talk as dead folks bein' better off dan dey was befo'."

Tom was silent, and Little Joe went on with renewed tears: "I come up to ax you to gimme a clean shirt an' a par o' draw's to put on Kiah. You needn't gimme no socks, 'cos he ain't got no feet. Oh, Lordy! oh, Lord!" sobbed Little Joe: "ef me an' Kiah had jes' had feet like some folks, Kiah wouldn't ha' been drownded!"

"Take this up to the house," said Mr. Wise, handing him a note, "and Miss Mollie will give you whatever you want."

"Thanky, sir," said Joe. "I know you ain't got no coffin handy, but you can gimme de money an' I can git one. I don't reckon it will take much, 'cos Kiah warn't big."

Then Mr. Wise wrote a note for the undertaker, and directed Joe what to do with it.

The next day was cold and dark and misty, and the pauper's hearse that conveyed Kiah to the graveyard was driven so fast that poor Little Joe, the only mourner, could hardly keep up as he hopped along behind it on his crutches.

The blast grew keener and the mist heavier, and before Kiah was buried out of sight, the rain was falling in torrents that drenched the poor little cripple sobbing beside the grave, and the driver of the hearse, a good-hearted Irishman, said to him, " In wid ye, or get up here by me, an' ye're a mind to. I'll take ye back."

But Joe shook his head, and prepared to hop back as he had hopped out. " Thanky, sir," said he, " but I'd ruther walk. I feels like I would be gittin' a ride out o' Kiah's funeral."

The wind blew open his buttonless shirt, and the rain beat heavily on his loyal little breast, but he struggled against the storm, and paused only once on his way home. That was beside the goods-box that he and Kiah had had for a stall. Now it was drenched with rain and the sides bespattered with mud, and the newspaper that had served for a cloth had blown over one corner and was soaked and torn, but clung to its old companion, though the wind tried to tear it away and the rain to beat it down. Little Joe stood a minute beside it, and cried harder than ever.

For several days Little Joe drooped and shivered and refused to eat, and at length he grew ill and sent for Mr. Wise ; but Mr. Wise was out of town, and did not return for a week ; and though, when he got home, the first thing he did was to visit Little Joe, he came too late, for Joe would never again rise from the straw pallet on which he lay, nor use the crutches that now stood idle in the corner.

His eyes brightened and he smiled faintly as Tom entered like a breath of fresh air,—so strong and fresh and vigorous that it made one feel better only to be near him.

"Why, Joe! how is this?"

The little cripple paused to gather up his strength; then he said, "Busted ag'in, Mas' Tom, and you can't nuvvur sot me up no mo'."

"Oh, stuff! Dr. North can if I can't. Why didn't you send for him when you found I was away?"

"I dunno, sir: I nuvvur thought 'bout it."

Turning to the woman with whom Joe lived, "And why didn't *you* do it?" said Tom angrily.

"I didn't know Joe was so sick," said she. "'Tain't no use sen'in' for no doctor now. I jes' been tellin' Joe he better not put off makin' peace wid de Lord."

"I don't reckon de Lord is mad wid me, Nancy. What is I done to Him? I didn't use to cuss, an' I didn't play marbles on Sunday, 'cos I couldn't play 'em *no* time, like de boys dat had feet."

"Ef you don't take keer you'll be too late, like Kiah. I ain't a-sayin' whar Kiah is now,— 'tain't for me to jedge," said Nancy,—"but you better be a-tryin' to open de gate o' Paradise."

Piping the words out slowly and painfully, Little Joe replied, "I don't b'leeve I keer 'bout goin' 'less Kiah can git in too; but I spec' he's dar, 'cos I don't see what de good Lord could ha' had ag'in' him. He oughtn't to thought hard o' nothin Kiah done, 'cos he warn't nuvvur nothin' but a free nigger, an' didn't hav no ole mas' to

pattern by. Maybe He'll let us bofe in. I know
Kiah's waitin' for me somewhar, but I dunno
what to say to Him. You ax Him, Mas' Tom."

He spoke more feebly, and his eyes were getting dull, but the old instinct of servitude remained, and he added, "Ain't you got nothin' to
spread on de flo', Nancy, so Mas' Tom won't git
his knees dirty?"

Immediately and reverently Tom knelt on the
clay floor, and, as nearly as he remembered it,
repeated the Lord's Prayer.

"Thanky, Mas' Tom," said Little Joe feebly.
"What was dat—ole mis'—used to—sing? 'Oh,
Lamb o'—God—I come—I—'" The words ceased,
and the eyes remained half closed, the pupils
fixed.

Little Joe was dead.

Jennie Woodville, in "Lippincott's Magazine."

THE MERCHANT AND THE BOOK-AGENT.

A BOOK-AGENT importuned James Watson, a
rich merchant living a few miles out of the city,
until he bought a book,—the "Early Christian
Martyrs." Mr. Watson didn't want the book,
but he bought it to get rid of the agent; then,
taking it under his arm, he started for the train
which takes him to his office in the city.

Mr. Watson hadn't been gone long before Mrs.
Watson came home from a neighbor's. The
book-agent saw her, and went in and persuaded
the wife to buy a copy of the book. She was

ignorant of the fact that her husband had bought
the same book in the morning. When Mr. Watson came back in the evening, he met his wife
with a cheery smile as he said, "Well, my dear,
how have you enjoyed yourself to-day? Well,
I hope?"

"Oh, yes! had an early caller this morning."

"Ah, and who was she?"

"It wasn't a 'she' at all; it was a gentleman,
a book-agent."

"A what?"

"A book-agent; and to get rid of his impor-
tuning I bought his book,—the 'Early Christian
Martyrs.' See, here it is," she exclaimed, advancing towards her husband.

"I don't want to see it," said Watson, frowning terribly.

"Why, husband?" asked his wife.

"Because that rascally book-agent sold me
the same book this morning. Now we've got
two copies of the same book,—two copies of the
'Early Christian Martyrs,' and—"

"But, husband, we can—"

"No, we can't, either!" interrupted Mr. Watson. "The man is off on the train before this.
Confound it! I could kill the fellow. I—"

"Why, there he goes to the depot now," said
Mrs. Watson, pointing out of the window at the
retreating form of the book-agent making for the
train.

"But it's too late to catch him, and I'm not
dressed. I've taken off my boots, and—"

Just then Mr. Stevens, a neighbor of Mr. Watson, drove by, when Mr. Watson pounded on the window-pane in a frantic manner, almost frightening the horse.

"Here, Stevens!" he shouted, "you're hitched up! Won't you run your horse down to the train and hold that book-agent till I come? Run! Catch 'im now!"

"All right," said Mr. Stevens, whipping up his horse and tearing down the road.

Mr. Stevens reached the train just as the conductor shouted, "All aboard!"

"Book-agent!" he yelled, as the book-agent stepped on the train. "Book-agent! hold on! Mr. Watson wants to see you."

"Watson? Watson wants to see me?" repeated the seemingly puzzled book-agent. "Oh, I know what he wants: he wants to buy one of my books; but I can't miss the train to sell it to him."

"If that is all he wants, I can pay for it and take it back to him. How much is it?"

"Two dollars, for the 'Early Christian Martyrs'" said the book-agent, as he reached for the money and passed the book out the car-window.

Just then Mr. Watson arrived, puffing and blowing, in his shirt-sleeves. As he saw the train pull out he was too full for utterance.

"Well, I got it for you," said Stevens,—"just got it, and that's all."

"Got what?" yelled Watson.

"Why, I got the book,—'Early Christian Martyrs,'—and paid—"

"By—the—great—guns!" moaned Watson, as he placed his hand to his brow and swooned right in the middle of the street.

BRER RABBIT AND THE TAR-BABY.

ONE evening, recently, the lady whom Uncle Remus calls "Miss Sally" missed her little seven-year-old. Making search for him through the house and through the yard, she heard the sound of voices in the old man's cabin, and, looking through the window, saw the child sitting by Uncle Remus. His head rested against the old man's arm, and he was gazing with an expression of the most intense interest into the rough, weather-beaten face that beamed so kindly upon him. This is what "Miss Sally" heard:

"Bimeby, one day, arter Brer Fox bin doin' all dat he could fer ter ketch Brer Rabbit, en Brer Rabbit bin doin' all he could fer ter keep 'im fum it, Brer Fox say to hisse'f dat he'd put up a game on Brer Rabbit, en he ain't mo'n got de wuds outen his mouf twel Brer Rabbit come a-lopin' up de big road, lookin' des ez plump en ez fat en ez sassy ez a Moggin hoss in a barley-patch.

"'Hol' on dar, Brer Rabbit,' sez Brer Fox, sezee.

"'I ain't got time, Brer Fox,' sez Brer Rabbit, sezee, sorter mendin' his licks.

"'I wanter have some confab wid you, Brer Rabbit,' sez Brer Fox, sezee.

"'All right, Brer Fox, but you better holler fum whar you stan': I'm monstus full er fleas dis mawnin',' sez Brer Rabbit, sezee.

"'I seed Brer B'ar yistiddy,' sez Brer Fox, sezee, 'en he sorter raked me over de coals kase you en me ain't make frens en live naberly, en I told him dat I'd see you.'

"Den Brer Rabbit scratch one year wid his off hine-foot sorter jub'usly, en den he ups en sez, sezee,—

"'All a-settin', Brer Fox. S'posen you drap roun' ter-morrer en take dinner wid me. We ain't got no great doin's at our house, but I speck de ole 'oman en de chilluns kin sort o' scramble roun' en git up sump'n fer ter stay yo' stummuck.'

"'I'm 'gree'ble, Brer Rabbit,' sez Brer Fox, sezee.

"'Den I'll 'pen' on you,' says Brer Rabbit, sezee.

"Nex' day, Mr. Rabbit an' Miss Rabbit got up soon, 'fo' day, en raided on a gyarden like Miss Sally's out dar, en got some cabbiges, en some roas'n-years, en some sparrergrass, en dey fix up a smashin' dinner. Bimeby one er de little Rabbits, playin' out in de back-yard, come runnin' in hollerin', 'Oh, ma! oh, ma! I seed Mr. Fox a-comin'!' En den Brer Rabbit he tuck de chilluns by der years en make um set down, and den him en Miss Rabbit sorter dally roun' waitin'

for Brer Fox. En dey keep on waitin', but no
Brer Fox ain't come. Atter 'while Brer Rabbit
goes to de do', easy like, en peep out, en dar,
stickin' out fum behime de cornder, wuz de tip-
een' er Brer Fox's tail. Den Brer Rabbit shot
de do' en sot down, en put his paws behime his
years, en begin fer to sing:

> " 'De place wharbouts you spill de grease,
> Right dar youer boun' ter slide,
> An' whar you fine a bunch er ha'r,
> You'll sholy fine de hide !'

"Nex' day Brer Fox sont word by Mr. Mink en
skuze hisse'f kase he wuz too sick fer ter come,
en he ax Brer Rabbit fer ter come en take dinner
wid him, en Brer Rabbit say he wuz 'gree'ble.

"Bimeby, w'en de shadders wuz at der shortes',
Brer Rabbit he sorter brush up en santer down
ter Brer Fox's house, en w'en he got dar he yer
somebody groanin', en he look in de do', en dar
he see Brer Fox settin' up in a rockin'-cheer all
wrop up wid flannil, en he look mighty weak.
Brer Rabbit look all 'roun', he did, but he ain't
see no dinner. De dish-pan wuz settin' on de
table, en close by wuz a kyarvin'-knife.

" 'Look like you gwineter have chicken fer dinner, Brer Fox,' sez Brer Rabbit, sezee.

" 'Yes, Brer Rabbit, deyer nice en fresh en
tender,' says Brer Fox, sezee.

"Den Brer Rabbit sorter pull his mustarsh, en
say, 'You ain't got no' calamus-root, is you, Brer
Fox? I done got so now dat I can't eat no
chicken 'ceppin' she's seasoned up wid calamus-

root.' En wid dat Brer Rabbit lipt out er de do'
and dodge 'mong de bushes, en sot dar watchin'
fer Brer Fox; en he ain't watch long, nudder,
kase Brer Fox flung off de flannil en crope out er
de house en got whar he could close in on Brer
Rabbit, en bimeby Brer Rabbit holler out, 'Oh,
Brer Fox! I'll des put yo' calamus-root out yer
on dis yer stump. Better come git it while hit's
fresh.' And wid dat Brer Rabbit gallop off
home. En Brer Fox ain't never kotch 'im yit, en
w'at's mo', honey, he ain't gwineter."

"Didn't the fox *never* catch the rabbit, Uncle
Remus?" asked the little boy the next evening.

"He come mighty nigh it, honey, sho's you
bawn,—Brer Fox did. One day atter Brer Rabbit fool 'im wid dat calamus-root, Brer Fox went
ter wuk en got 'im some tar, en mix it wid some
turkentime, en fix up a contrapshun what he call
a Tar-Baby, en he tuck dish yer Tar-Baby en he
sot 'er in de big road, en den he lay off in de
bushes fer ter see w'at de news wuz gwineter be.
En he didn't hatter wait long, nudder, kase bimeby here come Brer Rabbit pacin' down de road,
—lippity-clippity, clippity-lippity,—des ez sassy
ez a jay-bird. Brer Fox he lay low. Brer Rabbit come prancin' 'long twel he spy de Tar-Baby,
en den he fotch up on his behime legs like he
was 'stonished. De Tar-Baby she sot dar, she
did, en Brer Fox he lay low.

"'Mawnin'!' sez Brer Rabbit, sezee; 'nice
wedder dis mawnin',' sezee.

"Tar-Baby ain't sayin' nuthin', en Brer Fox he lay low.

"'How duz yo' sym'tums seem ter segashuate?'" sez Brer Rabbit, sezee.

"Brer Fox he wink his eye slow, en lay low, en de Tar-Baby she ain't sayin' nuthin'.

"'How you come on, den? Is you deaf?' sez Brer Rabbit, sezee. 'Kase if you is I kin holler louder,' sezee.

"Tar-Baby lay still, en Brer Fox he lay low.

"'Youer stuck up, dat's w'at you is,' says Brer Rabbit, sezee, 'en I'm gwineter kyore you, dat's w'at I'm a-gwineter do,' sezee.

"Brer Fox he sorter chuckle in his stummuck, he did, but Tar-Baby ain't sayin' nuthin'.

"'I'm gwineter larn you howter talk ter 'specttuble fokes ef hit's de las' ack,' sez Brer Rabbit, sezee. 'Ef you don't take off dat hat en tell me howdy, I'm gwineter bus' you wide open,' sezee.

"Tar-Baby stay still, en Brer Fox he lay low.

"Brer Rabbit keep on axin' 'im, en de Tar-Baby she keep on sayin' nuthin', twel present'y Brer Rabbit draw back wid his fis', he did, en blip he tuck er side er de head. Right dar's whar he broke his merlasses-jug. His fis' stuck, en he can't pull loose. De tar hilt him. But Tar-Baby she stay still, en Brer Fox he lay low.

"'Ef you don't lemme loose, I'll knock you ag'in,' sez Brer Rabbit, sezee; en wid dat he fotch 'er a wipe wid de udder han', en dat stuck. Tar-Baby she ain't sayin' nuthin', en Brer Fox he lay low.

"'Tu'n me loose 'fo' I kick de natal stuffin' outen you,' sez Brer Rabbit, sezee; but de Tar-Baby she ain't sayin' nuthin'. She des hilt on, en den Brer Rabbit lose de use er his feet in de same way. Brer Fox he lay low. Den Brer Rabbit squall out dat ef de Tar-Baby don't tu'n 'im loose he butt 'er crank-sided. En den he butted, en his head got stuck. Den Brer Fox he santered fort', lookin' des ez innercent ez wunner yo' mammy's mockin'-birds.

"'Howdy, Brer Rabbit?' sez Brer Fox, sezee. 'You look sorter stuck up dis mawnin',' sezee; en den he rolled on de groun', en laft en laft twel he couldn't laff no mo'. 'I speck you'll take dinner wid me dis time, Brer Rabbit. 1 done laid in some calamus-root, en I ain't gwineter take no skuse,' sez Brer Fox, sezee."

Here Uncle Remus paused, and drew a two-pound yam out of the ashes.

"Did the fox eat the rabbit?" asked the little boy to whom the story had been told.

"Dat's all de fur de tale goes," replied the old man. "He mout, en den ag'in he moutent. Some say Jedge B'ar come 'long en loosed 'im; some say he didn't. I hear Miss Sally callin'. You better run 'long." . . .

"Uncle Remus," said the little boy one evening, when he had found the old man with little or nothing to do, "did the fox kill and eat the rabbit when he caught him with the Tar-Baby?"

"Law, honey, ain't I tell you 'bout dat?" replied the old darky, chuckling slyly. "I 'clar ter grashus I ought er tole you dat; but ole man Nod wuz ridin' on my eyelids twel a lettle mo'n I'd 'a' dis'member'd my own name, en den on to dat here come yo' mammy hollerin' atter you.

"W'at I tell you w'en I fus' begin? I tole you Brer Rabbit wuz a monstus soon beas'; leas'ways dat's w'at I laid out fer ter tell you. Well, den, honey, don't you go en make no udder kalkala-shuns, kase in dem days Brer Rabbit en his family wuz at de head er de gang w'en enny racket wuz on han', en dar dey stayed. 'Fo' you begins fer ter wipe yo' eyes 'bout Brer Rabbit, you wait en see whar'bouts Brer Rabbit gwineter fetch up at. But dat's needer yer ner dar.

"W'en Brer Fox fine Brer Rabbit mixt up wid de Tar-Baby, he feel mighty good, en he roll on de groun' en laff. Bimeby he up 'n' say, sezee, —

"'Well, I speck I got you dis time, Brer Rab-bit,' sezee; 'maybe I ain't, but I speck I is. You been runnin' roun' here sassin' atter me a mighty long time, but I speck you done come ter de een' er de row. You bin cuttin' up yo' capers en bouncin' roun' in dis naberhood ontwel you come ter b'leeve yo'se'f de boss er de whole gang. En den youer allers some'rs whar you got no bizness,' sez Brer Fox, sezee. 'Who ax you fer ter come en strike up a 'quaintence wid dish yer Tar-Baby? En who stuck you up dar whar you is? Nobody in de roun' worril. You des tuck en jam yo'se'f on dat Tar-Baby widout waitin' fer enny invite,' sez Brer Fox, sezee, 'en dar you is, en dar you'll stay twel I fixes up a bresh-pile and fires her up, kase I'm gwineter bobbycue you dis day sho,' sez Brer Fox, sezee.

"Den Brer Rabbit talk mighty 'umble.

"'I don't keer w'at you do wid me, Brer Fox,' sezee, 'so you don't fling me in dat brier-patch. Roas' me, Brer Fox,' sezee, 'but don't fling me in dat brier-patch,' sezee.

"'Hit's so much trouble fer ter kindle a fire,' sez Brer Fox, sezee, 'dat I speck I'll hatter hang you,' sezee.

"'Hang me des ez high ez you please, Brer Fox," sez Brer Rabbit, sezee, 'but do fer de Lord's sake don't fling me in dat brier-patch,' sezee.

"'I ain't got no string,' sez Brer Fox, sezee, 'en now I speck I'll hatter drown you,' sezee.

"'Drown me ez deep ez you please, Brer Fox,' sez Brer Rabbit, sezee, 'but do don't fling me in dat brier-patch,' sezee.

"'Dey ain't no water nigh,' sez Brer Fox, sezee, 'en now I speck I'll hatter skin you,' sezee.

"'Skin me, Brer Fox,' sez Brer Rabbit, sezee, 'snatch out my eyeballs, t'ar out my years by de roots, en cut off my legs,' sezee, 'but do please, Brer Fox, don't fling me in dat brier-patch,' sezee.

"Co'se Brer Fox wanter hurt Brer Rabbit bad ez he kin, so he cotch him by de behime legs en slung 'im right in de middle er de brier-patch. Dar wuz a considerbul flutter whar Brer Rabbit struck de bushes, en Brer Fox sorter hang roun' fer ter see what wuz gwineter happen. Bimeby he hear somebody call 'im, en way up de hill he see Brer Rabbit settin' cross-legged on a chinka- pin log koamin' de pitch outen his ha'r wid a chip. Den Brer Fox know dat he bin swop off mighty bad. Brer Rabbit wuz bleedzed fer ter fling back some er his sass, en he holler out,—

"'Bred en bawn in a brier-patch, Brer Fox,—bred en bawn in a brier-patch!' en wid dat he skip out des ez lively ez a cricket in de embers."

Joel Chandler Harris.

A BATTLE OF WORDS.

WHEN Daniel O'Connell was yet a very young man, his talent for vituperative language was so great that he was deemed matchless as a scold. There lived in Dublin a certain woman, Biddy Moriarty by name, who kept a huckster's stall on one of the quays nearly opposite the Four Courts. She was a first-class virago,—formidable with both fist and tongue,—so that her voluble imputation had become almost proverbial in the country round about.

Some of O'Connell's friends thought that he could defeat her with her own weapons, while others ridiculed the idea. The Kerry barrister could not stand this, so he backed himself for a match. Bets were offered, and taken, and it was decided that the matter should be settled at once. So, proceeding to the huckster's stall with a few friends, O'Connell commenced his attack on the old lady :

"What is the price of this walking-stick, Mrs. What's-your-name?"

"Moriarty, sir, is my name, and a good one it is ; and what have you to say ag'in' it? and one and sixpence's the price of the stick. Troth it's cheap as dirt, so it is."

"One and sixpence for a walking-stick! whew! Why, you are no better than an impostor to ask eighteenpence for what cost you twopence."

"Twopence your grandmother!" replied Mrs. Biddy. "Do you mane to say that it's chating the people I am? Impostor, indeed!"

"Ay, impostor; and it's that I call you to your teeth," rejoined O'Connell.

"Come, cut your stick, you cantankerous jackanapes."

"Keep a civil tongue in your head, you old diagonal," cried O'Connell calmly.

"Stop your jaw, you pug-nosed badger, or by this and that," cried Mrs. Moriarty, "I'll make you go quicker nor you came."

"Don't be in a passion, my old radius: anger will only wrinkle your beauty."

"By the hokey, if you say another word of impudence, I'll tan your dirty hide, you bastely common scrub ; and sorry I'd be to soil my fists upon your carcass."

"Whew, boys! what a passion old Biddy is in! I protest, as I am a gentleman—"

"Jintleman! jintleman! the likes of you a jin-tleman! Wisha, by gor, that bangs Banagar! Why, you potato-faced pippinsneezer, when did a Madagascar monkey like you pick up enough of common Christian decency to hide your Kerry brogue?"

"Easy now,—easy now," cried O'Connell, with imperturbable good humor; "don't choke your-self with fine language, you old whiskey-drinking parallelogram."

"What's that you call me, you murderin' vil-lain?" roared Mrs. Moriarty, stung into fury.

"I call you," answered O'Connell, "a parallel-ogram; and a Dublin judge and jury will say it's no libel to call you so."

"Oh, tare-an-ouns! oh, holy Biddy! that an honest woman like me should be called a parry-bellygrum to her face! I'm none of your parry-bellygrums, you rascally gallows-bird, you cow-ardly, sneaking, plate-lickin' bliggard!"

"Oh, not you, indeed!" retorted O'Connell. "Why, I suppose you'll deny that you keep a hypothenuse in your house."

"It's a lie for you, you bastely robber! I never had such a thing in my house, you swin-dling thief!"

"Why, sure all of your neighbors know very well that you keep not only a hypothenuse, but that you have two diameters locked up in your garret, and that you go out to walk with them every Sunday, you heartless old heptagon."

" Oh, hear that, ye saints in glory ! Oh, there's bad language from a fellow that wants to pass for a jintleman! May the divil fly away with you, you wicher from Munster, and make celery-sauce of your rotten limbs, you mealy-mouthed bag of wind !"

" Ah, you can't deny the charge, you miserable sub-multiple of a duplicate ratio."

" Go rinse your mouth in the Liffey, you nasty tickle-pitcher; after all the bad words you are speakin' it ought to be filthier than your face, you dirty chicken of Beelzebub !"

" Rinse your own mouth, you wicked-minded polygon !—to the deuce I pitch you, you blustering intersection of a superficies !"

" You saucy tinker's apprentice, if you don't cease your jaw, I'll—" But here she gasped for breath, unable to hawk up any more words, for the last volley of O'Connell had nearly knocked the wind out of her.

" While I have a tongue I'll abuse you, you most inimitable periphery. Look at her, boys! there she stands,—a convicted perpendicular in petticoats. There's contamination in her circumference, and she trembles with guilt down to the extremities of her corollaries. Ah! you're found out, you rectilinear-antecedent and equiangular old hag! 'Tis with you the devil will fly away, you porter-swiping similitude of the bisection of a vortex !"

Overwhelmed with this torrent of language, Mrs. Moriarty was silenced. Catching up a saucepan, she was aiming at O'Connell's head when he very prudently made a timely retreat.

"You have won the wager, O'Connell: here's your bet," cried the gentleman who had proposed the contest.

Richard R. Madden.

SAM WELLER'S VALENTINE.

"Wot's that you're a doin' of?—pursuit of knowledge under difficulties?—eh, Sammy?"

"I've done now," said Sam, with slight embarrassment. "I've been a-writin'."

"So I see," replied Mr. Weller. "Not to any young 'ooman, I hope, Sammy?"

"Why, it's no use a-sayin' it ain't," replied Sam. "It's a walentine."

"A what?" exclaimed Mr. Weller, apparently horror-stricken by the word.

"A walentine," replied Sam.

"Samivel, Samivel!" said Mr. Weller, in reproachful accents, "I didn't think you'd ha' done it. Arter the warnin' you've had o' your father's wicious perpensities, arter all I've said to you upon this here wery subject, arter activally seein' and bein' in the company o' your own mother-in-law, vich I should ha' thought was a moral lesson as no man could ever ha' forgotten to his dyin' day! I didn't think you'd ha' done it, Sammy! I didn't think you'd ha' done it." These reflections were too much for the good old man. He raised Sam's tumbler to his lips and drank off its contents.

" Wot's the matter now ?" said Sam.

" Never mind, Sammy," replied Mr. Weller: "it'll be a wery agonizin' trial to me at my time of life, but I'm pretty tough, that's vun consolation, as the wery old turkey remarked ven the farmer said he was afeerd he should be obliged to kill him for the London market."

" Wot'll be a trial?" inquired Sam.

" To see you married, Sammy,—to see you a deluded wictim, and thinkin' in your innocence that it's all wery capital," replied Mr. Weller. " It's a dreadful trial to a father's feelin's, that 'ere, Sammy."

" Nonsense!" said Sam. " I ain't a-goin' to get married, don't you fret yourself about that : I know you're a judge o' these things. Order in your pipe, and I'll read you the letter— there."

We cannot distinctly say whether it was the prospect of the pipe, or the consolatory reflection that a fatal disposition to get married ran in the family and couldn't be helped, which calmed Mr. Weller's feelings and caused his grief to subside. We should be rather disposed to say that the result was attained by combining the two sources of consolation, for he repeated the second in a low tone very frequently, ringing the bell, meantime, to order in the first. He then divested himself of his upper coat, and, lighting the pipe and placing himself in front of the fire with his back towards it, so that he could feel its full heat and recline against the mantelpiece at the same time, turned towards Sam, and, with a countenance greatly mollified by the softening influence of tobacco, requested him to "fire away."

Sam dipped his pen into the ink to be ready for any corrections, and began with a very theatrical air :

"Lovely—"

"Stop," said Mr. Weller, ringing the bell. "A double glass o' the inwariable, my dear."

"Very well, sir," replied the girl, who with great quickness appeared, vanished, returned, and disappeared.

"They seem to know your ways here," observed Sam.

"Yes," replied the father : "I've been here before in my time. Go on, Sammy."

"'Lovely creetur,' " repeated Sam.

"'Tain't in poetry, is it?" interposed the father.

"No, no," replied Sam.

"Wery glad to hear it," said Mr. Weller. "Poetry's unnat'ral : no man ever talked in poetry, 'cept a beadle on boxin'-day, or Warren's blackin', or Rowland's oil, or some o' them low fellows. Never you let yourself down to talk poetry, my boy. Begin again, Sammy."

Mr. Weller resumed his pipe with critical solemnity, and Sam once more commenced, and read as follows :

"'Lovely creetur i feel myself a d . . . ——' "

"That ain't proper," said Mr. Weller, taking his pipe from his mouth.

"No ; it ain't that," observed Sam, holding the letter up to the light ; "it's 'shamed :' there's a blot there. 'I feel myself ashamed—'"

"Wery good," said Mr. Weller. "Go on."

"'Feel myself ashamed and completely cir—' I forget wot this here word is," said Sam, scratching his head with the pen, in vain attempts to remember.

"Why don't you look at it, then?" inquired Mr. Weller.

"So I *am* a-lookin' at it," replied Sam, "but there's another blot. Here's a 'c,' and a 'i,' and a 'd.'"

"Circumwented, p'r'aps," suggested Mr. Weller.

"No, it ain't that," said Sam : "'circumscribed,' that's it."

"That ain't as good a word as circumwented,' Sammy," said Mr. Weller gravely.

"Think not?" said Sam.

"Nothin' like it," replied his father.

"But don't you think it means more ?" inquired Sam.

"Vell, p'r'aps it is a more tenderer word," said Mr. Weller, after a few moments' reflection. "Go on, Sammy."

"'Feel myself ashamed and completely circumscribed in a dressin' of you, for you *are* a nice gal, and nothin' but it.'"

"That's a wery pretty sentiment," said the

elder Mr. Weller, removing his pipe to make way for the remark.

"Yes, I think it is rayther good," observed Sam, highly flattered.

"Wot I like in that 'ere style of writin'," said the elder Mr. Weller, "is, that there ain't no callin' names in it,—no Wenuses, nor nothin' o' that kind. Wot's the good o' callin' a young 'ooman a Wenus or a angel, Sammy?"

"Ah! what, indeed?" replied Sam.

"You might jist as vell call her a griffin, or a unicorn, or a king's arms, at once, which is wery vell known to be a col-lection o' fabulous animals," added Mr. Weller.

"Just as well," replied Sam.

"Drive on, Sammy," said Mr. Weller.

Sam complied with the request, and proceeded as follows, his father continuing to smoke, with a mixed expression of wisdom and complacency which was particularly edifying.

"'Afore I see you I thought all women was alike.'"

"So they are," observed the elder Mr. Weller parenthetically.

"'But now,'" continued Sam, "'now I find what a reg'lar soft-headed, ink-red'lous turnip I must ha' been, for there ain't nobody like you, though I like you better than nothin' at all.' I thought it best to make that rayther strong," said Sam, looking up.

Mr. Weller nodded approvingly, and Sam resumed:

" 'So I take the privilidge of the day, Mary, my dear,—as the gen'lem'n in difficulties did ven he valked out of a Sunday,—to tell you that the first and only time I see you your likeness was took on my heart in much quicker time and brighter colors than ever a likeness was took by the profeel-macheen, (which p'r'aps you may have heerd on, Mary, my dear,) altho' it *does* finish a portrait and put the frame and glass on complete with a hook at the end to hang it up by and all in two minutes and a quarter.' "

" I am afeerd that werges on the poetical, Sammy," said Mr. Weller, dubiously.

" No, it don't," replied Sam, reading on very quickly, to avoid contesting the point.

" 'Except of me, Mary, my dear, as your wal-entine, and think over what I've said.—My dear Mary, I will now conclude.' That's all," said Sam.

" That's raythur a sudden pull-up, ain't it, Sammy?" inquired Mr. Weller.

" Not a bit on it," said Sam : " she'll vish there wos more, and that's the great art o' letter-writin'."

" Well," said Mr. Weller, " there's somethin' in that ; and I wish your mother-in-law 'ud only conduct her conwersation on the same gen-teel principle. Ain't you a-goin' to sign it ?"

" That's the difficulty," said Sam ; "I don't know what *to* sign it."

" Sign it—Veller," said the oldest surviving proprietor of that name.

" Won't do," said Sam. " Never sign a walentine with your own name."

" Sign it ' Pickvick,' then," said Mr. Weller: " it's a wery good name, and an easy one to spell."

" The wery thing," said Sam. "I *could* end with a werse ; what do you think ?"

" I don't like it, Sam," rejoined Mr. Weller. " I never know'd a respectable coachman as wrote poetry, 'cept one, as made an affectin' copy o' werses the night afore he wos hung for a highway robbery ; and *he* wos only a Cambervell man, so even that's no rule."

But Sam was not to be dissuaded from the poetical idea that had occurred to him, so he signed the letter—

" Your love-sick
Pickwick."

Dickens.

THE INDIANS AND THE MUSTARD.

A PARTY of Indians were being *fêted* on the occasion of their first introduction to the manners and customs of the " pale-faces." The stoicism of the red man is a well-known trait. From childhood these children of the forest are schooled to endure pain without wincing or crying, and to be equally undemonstrative in their emotions of joy. Any departure from this standard of manliness they regard as a contemptible weakness. The Indians of our story were true " braves," whom no new experience, either of pleasure or displeasure, could startle into any sign more expressive than a grunt, their countenances being uniformly grave and impassive. Behold them at the festal board. Everything is novel and strange, yet they give no token of surprise, and scorn to betray their sense of awkwardness even by so much as asking questions. They take what is offered them and gulp it down with stern and desperate gravity. To one of them a pot of mustard is handed. He helps himself liberally to the mild-looking mixture, and swallows a good spoonful of it. Spirit of the tornado! Fiend of the burning prairie! What is this molten fire, compared to which the "fire-water" of the trader is as bland as milk? The unhappy warrior struggled to conceal his agony; but, though he succeeded in avoiding any contortion of the features, the tears, to his unspeakable disgust, chased themselves in a stream down his dusky cheeks. What would he not have given for an opportunity of scalping the innocent occasion of his trouble!

Meanwhile, his discomfort had not escaped the keen eyes of an Indian who sat beside him. Nudging his tearful comrade, the latter inquired, in low, guttural accents, the cause of his emotion. Suppressing his rage, the other mildly answered that he was thinking of his honored father who had lately gone to the happy hunting-grounds. Whether this explanation was regarded by the questioner as perfectly satisfactory we have no means of knowing; he did not, however, press his inquiries any further, nor does he appear to have suspected that the contents of the little jar had had anything in particular to do with the doleful memories of his friend. Presently the mustard came to *him*. It was a compound all untried; but the warrior was a stranger to fear. He took the condiment without hesitation, and he swallowed it freely—just once. Ah!! Death and torments! Is he on fire? Will he die? He is not quite sure; but it requires all his strength to keep quiet. The blood mounts to his head, and the tears—ugh! that he should thus play the squaw before all the company!—rush from his bulging eyes. Indian No. 1 is an interested observer of this little incident. His eyes had been upon the mustard-pot, and he had quietly awaited developments. His turn had now come; his revenge was at hand. Nudging his inwardly-writhing neighbor, he asked, in mildest gutturals, "My brother, why do you weep?" To which the furious sufferer gently replied, "I was weeping to think that when your precious father went to the happy hunting-grounds what a pity it was he did not take you with him."

CAUDLE'S WEDDING-DAY.

CAUDLE, love, do you know what next Sunday is? *No?* You don't! Well, was there ever such a strange man! Can't you guess, darling? Next Sunday, dear? Think, love, a minute,—just think. What! and you don't know now? Ha!

If I hadn't a better memory than you I don't know how we should ever get on. Well, then, pet, shall I tell you, dear, what next Sunday is? Why, then, it's our wedding-day. What are you groaning at, Mr. Caudle? I don't see anything to groan at. If anybody should groan, I'm sure it isn't you. No: I rather think it's I who ought to groan!

Oh, dear! That's fourteen years ago. You were a very different man then, Mr. Caudle. What do you say? *And I was a very different woman?* Not at all; just the same. Oh, you needn't roll your head about on the pillow in that way: I say, just the same. Well, then, if I'm altered, whose fault is it? Not mine, I'm sure, —certainly not. Don't tell me that I couldn't talk at all then: I could talk just as well then as I can now; only then I hadn't the same cause. It's you have made me talk. What do you say? *You're very sorry for it?* Caudle, you do nothing but insult me.

Ha! You were a good-tempered, nice creature fourteen years ago, and would have done anything for me. Yes, yes, if a woman would be always cared for she should never marry. There's quite an end of the charm when she goes to church! We're all angels while you're courting us; but once married, how soon you pull our wings off! No, Mr. Caudle, I'm not talking nonsense; but the truth is, you like to hear nobody talk but yourself. Nobody ever tells me that I talk nonsense but you. Now, it's no use your

turning and turning about in that way; it's not a
bit of— What do you say? *You'll get up?* No,
you won't, Caudle; you'll not serve me that trick
again, for I've locked the door and hid the key.
There's no getting hold of you in daytime; but
here you can't leave me. You needn't groan, Mr.
Caudle.

Now, Caudle, dear, do let us talk comfortably.
After all, love, there's a good many folks who, I
dare say, don't get on half so well as we've done.
We've both our little tempers, perhaps, but you are
aggravating, you must own that, Caudle. Well,
never mind; we won't talk of it; I won't scold
you now. We'll talk of next Sunday, love. We
never have kept our wedding-day, and I think it
would be a nice day to have our friends. What
do you say? *They'd think it hypocrisy?* No
hypocrisy at all. I'm sure I try to be comfortable; and if ever a man was happy, you ought to
be. No, Caudle, no; it isn't nonsense to keep
wedding-days; it isn't a deception on the world,
and if it is, how many people do it! I'm sure it's
only a proper compliment that a man owes to his
wife. Look at the Winkles: don't they give a
dinner every year? Well, I know; and if they
do fight a little in the course of the twelvemonth,
that's nothing to do with it. They keep their
wedding-day, and their acquaintance have nothing to do with anything else.

As I say, Caudle, it's only a proper compliment a man owes to his wife to keep his wedding-
day. It is as much as to say to the whole world,

"There, if I had to marry again, my blessed wife's the only woman I'd choose!" Well, I see nothing to groan at, Mr. Caudle,—no, nor to sigh at, either; but I know what you mean; I'm sure, what would have become of you if you hadn't married as you have done—why, you'd have been a lost creature! I know it; I know your habits, Caudle; and—I don't like to say it—but you'd have been little better than a ragamuffin. Nice scrapes you'd have got into, I know, if you hadn't had me for a wife. The trouble I've had to keep you respectable!—and what's my thanks? Ha! I only wish you'd had some women!

But we won't quarrel, Caudle. No; you don't mean anything, I know. We'll have this little dinner, eh? Just a few friends? Now, don't say you don't care; that isn't the way to speak to a wife, and especially the wife I've been to you, Caudle. Well, you agree to the dinner, eh? Now, don't grunt, Mr. Caudle, but speak out. You'll keep your wedding-day? What? *If I'll let you go to sleep?* Ha, that's unmanly, Caudle; can't you say, "Yes," without anything else? I say—can't you say, "Yes"? There, bless you! I knew you would.

And now, Caudle, what shall we have for dinner? No, we won't talk of it to-morrow; we'll talk of it now, and then it will be off my mind. I should like something particular,—something out of the way,—just to show that we thought the day something. I should like—Mr. Caudle, you're not asleep? *What do I want?* Why, you

know I want to settle about the dinner. *Have
what I like?* No; as it is your fancy to keep the
day, it's only right that I should try to please
you. We never had one, Caudle, so what do you
think of a haunch of venison? What do you say?
Mutton will do? Ha! that shows what you think
of your wife. I dare say if it was any of your
club friends—any of your pot-house companions
—you'd have no objection to venison. I say
if— What do you mutter? *Let it be venison?*
Very well. And now about the fish. What do
you think of a nice turbot? No, Mr. Caudle,
brill won't do; it shall be turbot, or there shan't
be any fish at all. Oh, what a mean man you are,
Caudle! Shall it be turbot? *It shall?* And
now about—the soup. Now, Caudle, don't swear
at the soup in that manner : you know there must
be soup. Well, once in a way, and just to show
our friends how happy we've been, we'll have
some real turtle. *No, you won't ? you'll have nothing but mock?* Then, Mr. Caudle, you may sit at
the table by yourself. Mock-turtle on a wedding-
day! Was there ever such an insult? What do
you say? *Let it be real, then, for once?* Ha,
Caudle! as I say, you were a very different person
fourteen years ago.

And, Caudle, you look after the venison!
There's a place I know, somewhere in the city,
where you'll get it beautiful. You'll look at it?
You will? Very well.

And, now, who shall we invite? *Who I like?*
Now, you know, Caudle, that's nonsense ; because

I only like whom you like. I suppose the Pretty-
mans must come. But understand, Caudle, I
don't have *Miss* Prettyman: I am not going to
have my peace of mind destroyed under my own
roof : if she comes, I don't appear at the table.
What do you say ? *Very well ?* Very well be it,
then.

And now, Caudle, you'll not forget the venison ?
In the city, my dear ! You'll not forget the
venison ? A haunch, you know,—a nice haunch.
And you'll not forget the venison ?. (*A loud snore.*)
Bless me, if he ain't asleep ! Oh, the unfeeling
men !

Douglas Jerrold.

A MODEST WIT.

A SUPERCILIOUS nabob of the East,—
　　Haughty, being great,—purse-proud, being
　　　rich,—
A governor, or general, at the least,
　　I have forgotten which,—
Had in his family a humble youth,
　　Who went to England in his patron's suite,
An unassuming boy, and yet in truth
　　A lad of decent parts and good repute.

This youth had sense and spirit;
　　But yet, with all his sense,
　　Excessive diffidence
Obscured his merit.

One day at table, flushed with pride and wine,
  His honor, proudly free, severely merry,
Conceived it would be vastly fine
  To crack a joke upon his secretary.

"Young man," he said, "by what art, craft, or
    trade
  Did your good father gain a livelihood?"
"He was a saddler, sir," Modestus said,
  "And in his time was reckoned good."

"A saddler, eh! and taught you Greek,
  Instead of teaching you to sew!
Pray, why did not your father make
  A saddler, sir, of you?"

Each parasite then, as in duty bound,
  The joke applauded, and the laugh went round.
At length Modestus, bowing low,
  Said (craving pardon, if too free he made),
"Sir, by your leave, I fain would know
  Your father's trade!"

"My father's trade?   Bless me, that's too bad!
My father's trade?   Why, blockhead, are you
    mad?
My father, sir, did never stoop so low:
He was a gentleman, I'd have you know."

"Excuse the liberty I take,"
  Modestus said, with archness on his brow,
"Pray, why did not your father make
  A gentleman of you?"

HIRING A COOK.

In the morning the old gentleman received the
visits of sundry tradesmen, to whom he had given
orders for different articles of dress; and Wilson, who was fully installed in his high office,
presented for his approbation Monsieur Rissolle,
"without exception the best cook in the United
Kingdom."

The particular profession of this person, the
colonel, who understood very little French, was
for some time puzzled to find out; he heard a
vocabulary of dishes enumerated with grace and
fluency, he saw a remarkably gentlemanly-look-
ing man, his well-tied neckcloth, his well-
trimmed whiskers, his white kid gloves, his
glossy hat, his massive chain encircling his neck,
and protecting a repeating Breguet, all pronounc-
ing the man of ton ; and when he came really
to comprehend that the sweet-scented, ring-
fingered gentleman before him was willing to
dress a dinner on trial, for the purpose of dis-
playing his skill, he was thunderstruck.

"Do I mistake ?" said the colonel: "I really
beg pardon,—it is fifty-eight years since I learned
French,—am I speaking to—a " (and he hardly
dared to pronounce the word)—" cook ?"

"Oui, monsieur," said M. Rissolle ; "I believe
I have de first reputation in de profession; I live
four years wiz de Marqui de Chester, and je me
flatte dat, if I had not turn him off last months,

I should have superintend his cuisine at dis moment."

"Oh, you have discharged the marquis, sir?" said the colonel.

"Yes, mon colonel, I discharge him, because he cast affront upon me, insupportable to an artist of sentiment."

"Artist!" mentally ejaculated the colonel.

"Mon colonel, de marqui had de mauvais goût one day, when he had large partie to dine, to put salt into his soup, before all his compagnie."

"Indeed," said Arden; "and, may I ask, is that considered a crime, sir, in your code?"

"I don't know code," said the man. "Morue? —dat is salt enough without."

"I don't mean *that*, sir," said the colonel: "I ask, is it a crime for a gentleman to put salt into his soup?"

"Not a crime, mon colonel," said Rissolle, "but it would be de ruin of me, as cook, should it be known to de world: so I told his lordship I must leave him; that de butler had said dat he saw his lordship put de salt into de soup, which was to proclaim to de universe dat I did not know de propre quantité of salt required to season my soup."

"And you left his lordship for *that*?" inquired the astonished country gentleman.

"Oui, sir. His lordship gave me excellent character; I go afterward to live wid my Lord Trefoil, very good, respectable man, my lord, of good family, and very honest man, I believe;

but de king, one day made him his gouverneur in Ireland, and I found I could not live in dat Dublin."

"No?"

"No, mon colonel: it is fine city," said Rissolle,—"good place,—but dere is no Italian Opera."

"How shocking!" said Arden. "And you left his Excellency on *that* account?"

"Oui, mon colonel."

"Why, his Excellency managed to live there without an Italian Opera," said Arden.

"Yes, mon colonel, c'est vrai; but I presume he did not know dere was none when he took de place. I have de character from my lord, to state why I leave him."

Saying which, he produced a written character from Lord Trefoil, who, being a joker as well as a minister, had actually stated the fact related by the unconscious turnspit as the reason for their separation.

"And pray, sir," said the colonel, "what wages do you expect?"

"Wages! Je n'entend pas, mon colonel," answered Rissolle. "Do you mean de stipend, —de salarie?"

"As you please," said Arden.

"My Lord Trefoil," said Rissolle, "give to me seven hundred pound a year, my wine, and horse and tilbury, with small tigre for him."

"Small what, sir?" exclaimed the astonished colonel.

"Tigre," said Rissolle; "little man-boy, to hold de horse."

"Ah! said Arden, "seven hundred pounds a year, and a tiger!"

"Exclusive of de pâtisserie, mon colonel. I never touch dat départment; but I have de honor to recommend Jenkin, my sister's husband, for de pâtisserie, at five hundred pound and his wine. Oh, Jenkin is dog ship at dat, mon colonel."

"Oh! exclusive of pastry," said the colonel emphatically.

"Oui, mon colonel," said Rissolle.

"Which is to be contrived for five hundred pounds per annum additional. Why, sir, the rector of my parish, a clergyman and a gentleman, with an amiable wife and seven children, has but half the sum to live upon."

"Dat is hard," said Rissolle, shrugging up his shoulders.

"Hard?—it *is* hard, sir," said Arden; "and yet you will hear the men who pay their cooks seven hundred a year for dressing dinners get up in their places in Parliament, declaim against the exorbitant wealth of the Church of England, and tell the people that our clergy are overpaid."

"Poor clergie! Mon colonel," said the man, "I pity your clergie; but den you don't remember de science and experience dat it require to make an omelette soufflée."

"Sir!" said Arden. "Do you mean seriously

and gravely to ask me seven hundred pounds a year for your services ?"

"Oui, vraiment, mon colonel," said Rissolle, at the same moment gracefully taking snuff from a superb gold box.

"Why, sir, I can't stand this any longer," cried the irritated novice in the fashionable world. "Seven hundred pounds! Make it guineas, sir, and I'll be *your* cook for the rest of my life."

The noise of this annunciation, the sudden leap taken by Monsieur Rissolle, to avoid something more serious than words, which he anticipated from the irate colonel, brought Wilson into the room, who, equally terrified with his Gallic friend at the symptoms of violent anger which his master's countenance displayed, stood wondering at the animation of the scene; when Arden, whose rage at the nonchalance of Rissolle at first impeded his speech, uttered, with an emphasis not to be misunderstood,—

"Good-morning, sir. Seven hundred—"

What the rest of this address might have been it is impossible to say, for before it was concluded Rissolle had left the apartment, and Wilson closed the door.

Theodore Hook.

MME. EEF.

MONSIEUR ADAM was all alone in ze garden. He have plenty for eat and plenty for drink and ees very comfortable, but he 'ave not much clothes.

Von evening he lie down on ze ground for take a nap. In ze morning he wake viz pain in his side.

He say : "Oh, mon Dieu, vat ees ze mattair, eh? Ah! ees von rib gone! I shall take un promenade in ze open air. I shall feel bettaire."

He promenade. Mme. Eef she approach. It is ze first lady zat M. Adam have ever met; it ees Mme. Eef's first entree to society. They approach each other and both are very much attract. M. Adam, he say : "Madame, shall I 'ave ze plaisair for promenade viz you?"

Mme. Eef respond, " I shall be most happy," and they valk together.

Zey promenade under un arbre; un arbre viz ze pretty appel on it; ze pretty appel viz ze red streak.

Monsieur le Serpent he sit up in ze arbre. He 'ave pretty mask all over hees face—look like elegant gentilhomme.

Madame Eef she see Monsieur le Serpent viz ze pretty mask and ze appel viz ze red streak, and she ees very much attract.

Monsieur le Serpent he say, "Madame Eef, shall I 'ave ze plaisair for peek you un appel?"

Madame Eef she reach out her hand for take ze appel.

Monsieur Adam he say : "Hola! hola! voila! Vat you do, eh? Do you not know ees prohibit? You must not touch ze appel! If you eat ze appel you shall become like un Dieu—you shall know ze good from ze evil!"

Monsieur le Serpent he take un pinch of snuff. He say : "Monsieur Adam, ees prohibit for you. If you eat ze appel you shall become like un Dieu—you shall know ze good from ze evil.   But Madame Eef—Madame Eef—she cannot become more of a goddess zan she ees now."

And zat finish Madame Eef.

SERMON.

BRETHREN :

The words of my text are :

> " Old Mother Hubbard, she went to the cupboard,
> To get her poor dog a bone ;
> But when she got there, the cupboard was bare,
> And so the poor dog had none."

These beautiful words, dear friends, carry with them a solemn lesson.

I propose this evening to analyze their meaning, and to attempt to apply it, lofty as it may be, to our every-day life.

> "Old Mother Hubbard, she went to the cupboard,
> To get her poor dog a bone."

Mother Hubbard, you see, was old ; there being no mention of others, we may presume she was alone ; a widow—a friendless, old, solitary widow.

Yet, did she despair ? Did she sit down and weep, or read a novel, or wring her hands ? No ! *" she went to the cupboard."*   And here observe that she *went* to the cupboard.   She did not hop, or skip, or run, or jump, or use any other peripatetic artifice; she solely and merely *went* to the cupboard.

We have seen that she was old and lonely, and we now further see that she was poor. For, mark, the words are " *the* cupboard."

Not "one of the cupboards," or the "right-hand cupboard," or the "left-hand cupboard," or the one above, or the one below, or the one under the stair, but just *the* cupboard. The one little humble cupboard the poor widow possessed. And why did she go to the cupboard? Was it to bring forth golden goblets, or glittering precious stones, or costly apparel, or feasts, or any other attributes of wealth? *It was to get her poor dog a bone!* Not only was the widow poor, but her dog, the sole prop of her age, was poor too.

We can imagine the scene. The poor dog crouching in the corner, looking wistfully at the solitary cupboard, and the widow going to that cupboard—in hope, in expectation maybe—to open it, although we are not distinctly told that it was not half open, or ajar, to open it for that poor dog.

> " But when she got there, the cupboard was bare,
> And so the poor dog had none."

"When she got there!" You see, dear brethren, what perseverance is.

You see the beauty of persistence in doing right. *She got there.*

There were no turnings and twistings, no slippings and slidings, no leaning to the right, or falterings to the left.

With glorious simplicity we are told *she got there.*

And how was her noble effort rewarded?

"The cupboard was bare!" It was bare! There were to be found neither oranges, nor cheesecakes, nor penny buns, nor gingerbread, nor crackers, nor nuts, nor lucifer matches.

The cupboard was bare!

There was but one, only one, solitary cupboard in the whole of that cottage, and that one, the sole hope of the widow and the glorious loadstar of the poor dog, was bare! Had there been a leg of mutton, a loin of lamb, a fillet of veal, even an ice from Gunter's, the case would have been different, the incident would have been otherwise.

But it was bare, my brethren, bare as a bald head, bare as an infant born without a caul.

Many of you will probably say, with all the pride of worldly sophistry, The widow, no doubt, went out, and bought a dog-biscuit.

Ah, no! Far removed from these earthly ideas, these mundane desires, poor Mother Hubbard, the widow, whom many thoughtless worldlings would despise, in that she only owned one cupboard, perceived—or I might even say, saw—at once the relentless logic of the situation, and yielded to it with all the heroism of that nature which had enabled her without deviation to reach the barren cupboard.

She did not attempt, like the stiff-necked scoffers of this generation, to war against the inevitable ; she did not try, like the so-called men of science, to explain what she did not understand.

She did nothing. "The poor dog had none!" And then, at this point, our information ceases.

But do we not know sufficient? Are we not cognizant of enough ?

Who would dare to pierce the veil that shrouds the ulterior fate of old Mother Hubbard, the poor dog, the cupboard, or the bone that was not there ?

Must we imagine her still standing at the open cupboard door—depict to ourselves the dog still drooping his disappointed tail upon the floor— the sought-for bone still remaining somewhere else ?

, Ah, no! my dear brethren, we are not so permitted to attempt to read the future. Suffice it for us to glean from this beautiful story its many lessons ; suffice it for us to apply them, to study them as far as in us lies, and, bearing in mind the natural frailty of our nature, to avoid being widows ; to shun the patronymic of Hubbard ; to have, if our means afford it, more than one cupboard in the house, and to keep stores in them all.

And oh! dear friends, keeping in recollection what we have learned this day, let us avoid keeping dogs that are fond of bones.

But, brethren, if we do—if fate has ordained that we should do any of these things—let us then go, as Mother Hubbard did, straight, without curveting or prancing, to our cupboard, empty though it be; let us, like her, accept the inevitable with calm steadfastness; and should we, like her, ever be left with a hungry dog and an empty cupboard, may future chroniclers be able to write also of us, in the beautiful words of our text:

"And so the poor dog had none."

Portsmouth (*Eng.*) *Monitor.*

TEDDY'S SIX BULLS.

A MERRY evening party in an English country town were bantering poor Teddy O'Toole, the Irishman, about his countrymen being so famous for bulls.

"By my faith," said Teddy, "you needn't talk about that same in this place: you're as fond of bulls as any people in all the world, so you are."

"Nonsense!" some of the party replied; "how do you make that out?"

"Why, sure, it's very aisy, it is; for in this paltry bit of a town you've got more public houses nor I ever seen wid the sign of the bull over the doors, so you have," said Teddy.

"Nay, Teddy, very few of those; but there's some of 'em, you know, in every town."

"Yes," said Teddy, obstinately sticking to his text, for he had laid a trap for his friends; "but you've more nor your share, barring that you're so fond of bulls, as I say. I'm sure I can count half a dozen of 'em."

" Pooh, nonsense !" cried the party : " that will never do. What'll you bet on that, Teddy? You're out there, my boy, depend upon it : we know the town as well as you ; and what will you bet ?"

"Indeed, my brave boys, I'll not bet at all. I'm no better, I assure ye : I should be worse, if I wur." This sally tickled his companions, and he proceeded : " But I'll be bound to name and count the six."

"Well, do, do," said several voices.

" Now, let me see ; there's the Black Bull."

" Yes, that's one."

" Then, there's the Red Bull."

" That's two."

" And the White Bull."

" Come, that's three."

" And the Pied Bull."

" So there is ; you'll not go much farther."

" And then there's—there's—there's the Golden Bull, in—what's it street ?"

" Well done, Teddy ; there's five, sure enough ; but you're short yet."

" Ay," said the little letter-carrier, who sat smirking in the corner, " and he will be short ; for there isn't one more, I know."

" And then, remember," continued Teddy, carefully pursuing his enumeration, " there's the Dun Cow."

At this a burst of laughter fairly shook the room, and busy hands kept the tables and glasses rattling, amidst boisterous cries of,—

"A bull! a bull!"

Looking seriously at all around, Teddy deliberately asked,—

" Do you call that a bull?"

"To be sure, it's a bull," exclaimed several voices at once.

"Then," said Teddy, "that's the sixth."

A RAILWAY MATINEE.

THE last time I ran home over the Chicago, Burlington and Quincy we had a very small, but select and entertaining party on the train. It was a warm day, and everybody was tired with the long ride and oppressed by the heat. The precise woman, with her hat ~~annoyed in my immense blue veil,~~ who always parsed her sentences before she uttered them, utterly worn out and thoroughly lonesome, was glad to respond to the pleasant nod of the big rough man who got on at Monmouth, and didn't know enough grammar to ask for the mustard so that you could tell whether he wanted you to pass it to him or pour it on his hair. The thin, troubled-looking man with the sandy goatee, who stammered so dreadfully that he always forgot what he wanted to say before he got through wrestling with any word with a " W" in it, lit up with a tremulous, hesitating smile, as he noticed this indication of sociability, for, like most men who find it extremely difficult to talk at all, he wanted to talk all the time. And the fat old gentleman sitting opposite him, who was so deaf that he couldn't hear the cars rattle, and always awed and bothered the stammerer into silence by saying "Hey?" in a very imperative tone, every time he got in the middle of a hard word, cocked his irascible head on one side as he saw this smile, and after listening intently to dead silence for a minute, suddenly broke out with such an emphatic, impatient, "Hey?" that everybody in the car started up and shouted, nervously and ungrammatically: "I didn't say nothing!" with the exception of the woman with the blue veil, who said : "I said nothing."

The fat old gentleman was a little annoyed and startled by such a chorus of responses, and fixing his gaze still more intently upon the thin man, said defiantly :

"Wha' say ?"

"I-I-I-I w-w-wuh-wuh-wasn'-wasn'—I wasn' s-s-sp—speak—"

"Hey?" roared the fat man.

"He wa'n't sayin' nauthin'," shouted the big rough man, nodding friendly encouragement to the thin man ; "he hain't opened his mouth !"

"Soap in the South ?" queried the fat old gentleman, impatiently. "Wha' for ?"

"Mouth, mouth," explained the precise woman, with impressive nicety. "He said 'opened his mouth.' The gentleman, seated directly opposite you was—"

"Offers to chew what ?" cried the fat old gentleman in amazement.

"Sir," said the precise woman, "I made no

reference whatever to chewing. You certainly misunderstood me."

The thin man took courage from so many re-inforcements, and broke in :

" I-I-I-I d-d-d-dud-d-u-d-d-u-d-don't don't— I don't ch-ch-ch—"

" Hey ?" shouted the fat gentleman.

" He don't chaw nauthin' !" roared the big rough man, in a voice that made the car windows rattle. " He wa'n't a talkin' when you shot off at him !"

" Who got off ?" exclaimed the fat old gentle-man. " Wha'd' he get off for ?"

" You don't appear to comprehend clearly what he stated," shrieked the precise woman. " No person has left the train."

" Then wha'd' he say so for ?" shouted the fat man.

" Oh !" said the thin man, in a surprising burst of fluency ; " he-he-de-d-d-did did—"

" Who did ?" queried the fat man, talking louder than any one else.

" Num-num-num-num-n-no-nobody nobody. He he-d-d-d-d-dud-didn't didn't s—"

" Then wha' made you say he did ?" howled the deaf man.

" You misunderstand him," interrupted the precise woman. " He was probably about to re-mark that no reference whatever had been inten-tionally made to the departure of any person from the train, when you interrupted him in the midst of an unfinished sentence, and hence ob-

tained an erroneous impression of the tenor of
his remarks. He meant no offense—"

"Know a fence?" roared the fat man. "Of
course I know a fence!"

"He hain't got middlin' good hearin'," yelled
the big man, as apologetically as a steam whistle
could have shrieked it. "Y'ears kind of stuffed
up!"

"Time to brush up?" cried the fat man. "Wha'
for?"

"No," shrieked the precise woman; "he remarked to the other gentleman that your hearing
appeared to be rather defective."

"His father a detective?" hooted the fat gentleman, in amazement.

"N-n-n-n-nun-nun-no!" broke in the thin man;
"h-h-h-h-huh-huh-he-s-s-sa-sa-said-said you w-
w-w-wuh was a little dud-dud—was a little
deaf."

"Said I was a thief!" howled the fat man, a
scarlet tornado of wrath; "said I was a thief!
Wha'd'ye mean? Show him to me! Who says
I'm a thief? Who says so?"

"Now," shouted the big rough man, "nobody
don't say ye ain't no thief. I jest sayed as how
we didn't git along very well. Ye see he," nodding to the thin man, "he can't talk very well,
an'—"

"Wh-wh-wh-why c-c-can't I t-t-t-tut-tut-tut-
talk?" broke in the thin man, white with rage.
"I-I-I-I'd like t-t-to know wh-wh-wh-what's the
reason I c-c-can't tut-tut-talk as w-w-w-well as

any bub-bub-body that's bub-bub-bub-been tut-tut-talking on this car ever s-s-s-since the tut-tut-tut—"

"Hey?" roared the fat man, in an explosion of indignant suspicion.

"I was sayin'," howled the big rough man, "as how he didn't talk middlin' well—"

"Should say so," growled the fat man, in tones of intense satisfaction.

"An'," the big rough man went on, yelling with delight at having made the old party hear something, "an' you can't hear only tollable—"

"Can't hear?" the fat old gentleman broke out in a resonant roar. "Can't hear! Like to know why I can't hear! Why can't I? If I couldn't hear better than half the people on this train I'd cut off my ears! Can't hear? It's news to me if I can't. I'd like to know who—" "Burlington!" yelled the brakeman. "Chang' car f'r Keokuk, Ceed Rap's an' For' Mad'son! This car f'r Omaha! Twen' mints f'r supper!"

And but for this timely interruption, I don't think our pleasant little party would have got out of that snarl this side of San Francisco.

R. J. Burdette.

UNCLE REUBEN'S BAPTISM.

HE is an industrious colored man, living in a small cabin down the river; and his wife is a corpulent, good-natured woman, but very deaf.

Some weeks ago, Reuben began to ponder.

He had never been a bad darky; but he had never embraced Christianity, much to the sorrow of Aunt Susan, his wife, who has been prepared for heaven, lo, these many years past. The more he pondered, the more he became convinced that he ought to become a Christian; and Aunt Susan encouraged him with tender words and tearful eyes.

The old man came to town several days ago to see about joining a church, and was informed that he ⁄ould have to be baptized before he could become a member. He didn't relish the idea much; but he informed his wife that he would consent; and she clasped her hands, and replied,

"Glory to Richmond! De angels am a-comin'!"

Uncle Reuben got the idea, the other day, that he'd like to try the water alone, before being publicly baptized; and while his wife was getting breakfast ready, he slipped down to the riverbank to take a preparatory dip. He removed his coat, hat, and boots, placed them on a log, and as he descended the bank, his broad feet slipped, and the convert came down on the back of his neck.

"What de debbil!"—he commenced, as he picked himself up; but suddenly remembering that he was soon to join the church, he checked himself, and remarked,—

"I'm ashamed of dat; and I hope de angels will 'scuse me."

He put one foot into the water, and drew back with a shiver ; put in the other, and looked longingly toward the house. At that moment Aunt Susan began singing,—

> " We's gwine up to glory :
> We's gwine on de cars."

And old Reuben braced up, and entered the water.

" Yes ; we's gwine up to glory !" he remarked as he waded along,—" gwine on de fast express."

At the next step, his foot struck a sunken log : and he pitched over it, under water, head first. As soon as he came to the surface, and blew the water from his mouth, he yelled,—

" Woosh ! What in blazes is dis yere performance ?"

In raising up, his foot slid over the log, and under a limb, in such a manner, that the old darky was caught fast. He could hang to a stub of a limb, but he could not put himself forward enough to slip his foot out of the trap.

" Whar de angels now ?" he yelled out, as he kicked the water higher than his head.

Aunt Susan answered with,—

> " De angels am a-comin':
> I hear de music play."

When the old man realized that he was fast, and must have help from the shore, he yelled out,—

" Ho, dare, old woman ! Hi !"

She couldn't have heard a cannon fire on the bank of the river, and went on singing,—

" Dare's a seat for me in heaven :
   I'se gwine to jine de band."

" Hi, dare ! I'll jine your old black head off, if you don't hear me !" yelled old Reuben.

He struggled and kicked, got his head under water, and out, and yelled,—

"Cuss dat old woman ! Why don't she hear me ?"

" Uncle Reuben's a-gwine
   To be an angel sho',"

came the song.

" It's a lie, a big debbil lie !" he yelled, pulling his head under water again.

" And he'll fly among de angels,
   And play upon a harp,"

continued the old woman, as she turned over the bacon.

" Hi, dare ! woosh, whoop !" he yelled, floundering around, pulling at his leg.

"De Lawd has got his name,
   And dere is a place for him !"

howled the old woman.

" Whoa dare, you old black villum !" yelled Uncle Reuben.

"Dey'll dress him up in white,
   Wid a crown upon his brow,"

wailed Aunt Susan, as she poured the water off the potatoes.

"If I ebber git out o' dis ribber alive, I'll break her old deaf head, I will!" growled the victim ; and then, raising his voice, he shouted,—

"You dare, old Satan, hi, hi!" As if in direct answer, came the song,—

> "He struggles wid de evil one,
> But he gained de vict'ry, sho'!"

"Susan, Susan!" if I had you by de wool, I'd barry dat old deaf head agin de cabin till yer eyes couldn't see!" he screamed ; and he made another tremendous effort to get loose. It was successful ; and just then she sang,—

> "Oh! whar's de angel now ?
> Send him 'long; send him 'long!"

"De angel am a-comin'!" growled Uncle Reuben as he waded ashore ; "and he'll turn dat cabin inside out!"

He limped up to the house. She was placing the meal on the table, and singing,—

> "He's gwine to be baptized;
> He's gwine—"

when he entered the house, and gave her a cuff on the ear which nearly loosened the roots of her hair.

"Oh, yes! I'ze an angel wid wings on, I is!" he yelled, as he brought her another cuff; "and I'ze gwine to glory,—and I'll knock yer old head

off!—and I'ze gwine to jine de band—and you
deaf old alligator!—and I'ze gwine up to heaven
—and blame yer old deaf ears!—and de glory
am a-comin'!"

People who know Uncle Reuben say that he
swears again with great relish ; and it is certain
that he hasn't been up to Vicksburgh to be baptized and become a church-member.

THE YELLOW DOMINO.

In the latter part of the reign of Louis XV. of
France the masquerade was an entertainment
high in estimation, and was often given at an immense cost on court-days and such occasions of
rejoicing. As persons of all ranks might gain
admission to these spectacles, provided they
could afford the purchase of the ticket, very
strange rencontres frequently took place at them,
and exhibitions almost as curious in the way of
disguise or assumption of character. But perhaps the most whimsical among the genuine surprises recorded at any of these spectacles was
that which occurred in Paris the 15th of October,
on the day when the dauphin attained the age of
one-and-twenty.

At this fête, which was of a peculiarly glittering character,—so much so, that the details of it
are given at great length by the historians of the
day,—the strange demeanor of a man in a yellow domino, early in the evening, excited atten-

tion. This mask, who showed nothing remarkable as to figure, though tall rather, and of robust proportions, seemed to be gifted with an appetite not merely past human comprehension, but passing the fancies even of romance.

> " The dragon of old, who churches ate
> (He used to come on a Sunday),
> Whole congregations were to him
> But a dish of Salmagundi,"

was a nibbler—a mere fool—to this stranger of the yellow domino. He passed from chamber to chamber, from table to table of refreshments, not tasting, but devouring, devastating, all before him. At one board he despatched a fowl, two thirds of a ham, and half a dozen bottles of champagne ; the very next minute he was found seated in another apartment, performing the same feat with a stomach better than at first. This strange course went on until the company, who at first had been amused by it, became alarmed and tumultuous.

"Is it the same mask ? or are there several dressed alike ?" demanded an officer of the guards, as the yellow domino rose from a seat opposite to him and left the apartment.

"I have seen but one, and, by heavens, he is here again !" exclaimed the party to whom the query was addressed.

The yellow domino spoke not a word, but proceeded straight to the vacant seat which he had just left, and again commenced supping, as though he had fasted for the half of a campaign.

At length the confusion which the proceeding created became universal, and the cause reached the ears of the dauphin.

"He is a very fiend, your highness," exclaimed an old nobleman, "or wants but a tail to be so!"

"Say rather he should be a famished poet, by his appearance," replied the prince, laughing. "But there must be some juggling: he spills all his wine, and hides the provisions under his robe."

Even while they were speaking, the yellow domino entered the room in which they were talking, and, as usual, proceeded to the table of refreshments.

"See here, my lord," cried one; "I have seen him do this twice."

"I thrice!"—"I five times!"—"And I fifteen!"

This was too much. The master of ceremonies was questioned. He knew nothing, and the yellow domino was interrupted as he was carrying a bumper of claret to his lips.

"The prince desires that Monsieur who wears the yellow domino should unmask." The stranger hesitated.

"The command with which his highness honors Monsieur is perfectly absolute."

Against that which is absolute there is no contending. The yellow man threw off his mask and domino, and proved to be a private trooper of the Irish dragoons!

"And in the name of gluttony, my good friend (not to ask how you gained admission), how have you contrived," said the prince, " to sup to-night so many times ?"

" Sire, I was but beginning to sup, when your royal message interrupted me."

"Beginning!" exclaimed the dauphin, in amazement. " Then what is it I have heard and seen? Where are the herds of oxen that have disappeared and the hampers of burgundy ! I insist upon knowing how this is !"

" It is, sire," returned the soldier, " may it please your grace, that the troop to which I belong is to-day on guard. We have purchased one ticket among us, and provided this yellow domino, which fits us all. By which means the whole of the front rank, being myself the last man, have supped, if the truth must be told, at discretion; and the leader of the second rank, saving your highness's commands, is now waiting outside the door to take his turn."

MISS EDITH HELPS THINGS ALONG.

My sister 'll be down in a minute, and says
    you're to wait, if you please,
And says I might stay 'til she came, if I'd prom-
    ise her never to tease,
Nor speak 'til you spoke to me first. But that's
    nonsense, for how would you know
What she told me to say, if I didn't? Don't you
    really and truly think so?

And then you'd feel strange here alone! And
    you wouldn't know just where to sit:
For that chair isn't strong on its legs, and we
    never use it a bit.
We keep it to match with the sofa. But Jack
    says it would be like you
To flop yourself right down upon it and knock
    out the very last screw.

S'pose you try? I won't tell. You're afraid to! O!
    you're afraid they would think it was mean!
Well, then, there's the album—that's pretty, if
    you're sure that your fingers are clean,
For sister says sometimes I daub it; but she only
    says that when she's cross.
There's her picture. You know it? It's like her;
    but she ain't as good-looking, of course!

This is me. It's the best of 'em all. Now, tell
    me, you'd never have thought
That once I was little as that? It's the only one
    that could be bought:
For that was the message to pa from the photo-
    graph man where I sat—
That he wouldn't print off any more till he first
    got his money for that.

What? Maybe you're tired of waiting. Why,
    often she's longer than this.
There's all her back hair to do up and all of her
    front curls to friz.

But it's nice to be sitting here talking like grown
      people, just you and me.
Do you think you'll be coming here often? Oh,
      do! But don't come like Tom Lee.

Tom Lee. Her last beau. Why, my goodness!
      he used to be here day and night,
Till the folks thought he'd be her husband ; and
      Jack says that gave him a fright.
You won't run away, then, as he did? for you're
      not a rich man, they say.
Pa says you are poor as a church mouse. Now,
      are you? And how poor are they ?

Ain't you glad that you met me? Well, I am ;
      for I know now that your hair isn't red.
But what there is left of it's mousy, and not what
      that naughty Jack said.
But there! I must go. Sister's coming. But I
      wish I could wait, just to see
If she ran up to you to greet you in the way that
      she used to greet Lee."

B. Harte.

THE BACHELOR'S DREAM.

My pipe is lit, my grog is mixed,
  My curtains drawn and all is snug,
Old Puss is in her elbow-chair,
  And Tray is sitting on the rug.

Last night I had a curious dream,
  Miss Susan Bates was Mistress Mogg—
What d'ye think of that, my Cat?
  What d'ye think of that, my Dog?

She looked so fair, she sang so well,
  I could but woo and she was won.
Myself in blue, the bride in white,
  The ring was placed, the deed was done!
Away we went in chaise-and-four,
  As fast as grinning boys could flog—
What d'ye think of that, my Cat?
  What d'ye think of that, my Dog?

What loving tête-à-têtes to come!
  But tête-à-têtes must still defer!
When Susan came to live with me,
  Her mother came to live with her!
With sister Belle she couldn't part,
  But all *my* ties had leave to jog—
What d'ye think of that, my Cat?
  What d'ye think of that, my Dog?

The mother bought a pretty poll,
  A monkey too,—what work he made!
The sister introduced a beau,
  My Susan brought a favorite maid.
She had a Tabby of her own,—
  A snappish mongrel christened Gog—
What d'ye think of that, my Cat?
  What d'ye think of that, my Dog?

The monkey bit, the parrot screamed,
  All day the sister strummed and sung,
The petted maid was such a scold!
  My Susan learned to use her tongue;
Her mother had such wretched health,
  She sate and croaked like any frog—
What d'ye think of that, my Cat?
  What d'ye think of that, my Dog?

No longer Deary, Duck, and Love,
  I soon came down to simple "M!"
The very servants crossed my wish,
  My Susan let me down to them.
The poker hardly seemed my own,
  I might as well have been a log—
What d'ye think of that, my Cat?
  What d'ye think of that, my Dog?

My clothes they were the queerest shape!
  Such coats and hats she never met!
My ways they were the oddest ways!
  My friends were such a vulgar set!
Poor Tompkinson was snubbed and huffed,
  She could not bear that Mister Blogg—
What d'ye think of that, my Cat?
  What d'ye think of that, my Dog?

At times we had a spar, and then
  Mamma must mingle in the song;
The sister took a sister's part,
  The maid declared her master wrong,

The parrot learned to call me "Fool!"
  My life was like a London fog—
What d'ye think of that, my Cat?
  What d'ye think of that, my Dog?

My Susan's taste was superfine,
  As proved by bills that had no end—
I never had a decent coat,
  I never had a coin to spend!
She forced me to resign my club,
  Lay down my pipe, retrench my grog—
What d'ye think of that, my Cat?
  What d'ye think of that, my Dog?

Each Sunday night we gave a rout
  To fops and flirts, a pretty list;
And when I tried to steal away,
  I found my study full of whist!
Then, first to come and last to go,
  There always was a Captain Hogg—
What d'ye think of that, my Cat?
  What d'ye think of that, my Dog?

Now was that not an awful dream
  For one who single is and snug—
With Pussy in the elbow-chair
  And Tray reposing on the rug?—
If I must totter down the hill,
  'Tis safest done without a clog—
What d'ye think, my Cat?
  What d'ye think, my Dog?

Hood.

HALF-WAY DOIN'S.

BELUBBED fellow-trabellers, in holdin' forth to-
    day,
I doesn't quote no special verse for what I has to
    say ;
De sermon will be berry short, an' dis here am de
    tex' :
Dat *half-way* doin's ain't no 'count in dis word'
    nor de nex'.
Dis worl' dat we's a-libbin' in is like a cotton row,
Where ebery cullud gentleman has got his line to
    hoe ;
An' ebery time a lazy nigger stops to take a nap,
De grass keeps on a-growin' for to smudder up
    de crap.

When Moses led de Jews acrost de waters of de
    sea,
Dey had to keep a-goin' jus' as fas' as fas' could
    be ;
Do you suppose dey could eber hab succeeded in
    dere wish,
And reached de promised land at last, if they had
    stopped to fish?
My frien's, dere was a garden once, where Adam
    libbed wid Eve,
Wid no one roun' to bodder dem, no nabors for
    to thieve ;
An ebery day was Christmas, an' dey had dere
    rations free,
An' eberyting belonged to dem except an apple-
    tree.

You all know 'bout de story,—how de snake come
    snookin' 'round,
A stump-tail, rusty moccasin, a-crawlin' on de
    ground,
How Eve an' Adam ate de fruit, an' went an' hid
    dere face,
Till de angel oberseer came an' drove dem off de
    place.
Now, s'pose dis man an' 'ooman, too, hadn't
    'tempted for to shirk,
But had gone about dere gardenin', an' 'tended
    to dere work,
Dey wouldn't have been loafin' where dey had no
    business to,
An' de debble nebber'd got a chance to tell 'em
    what to do.

No *half-way doin's*, bredren, 'twill nebber do, I
    say !
Go at your task, an' finish it, an' den's de time to
    play ;
For even if de crap is good, de rain will spoil de
    bolls,
Unless you keeps a-pickin' in de garden ob your
    souls.
Keep a-ploughin', an' a-hoein', an' a-scrapin' up
    de rows ;
An' when de ginnin's ober you can pay up what
    you owes ;
But if you quits a-workin' ebery time de sun is
    hot
De sheriff's gwine to leby upon eberyting you's
    got.

Whateber you's a-dribin' at, be sure an' dribe it
    t'ro',
An' don't let nothin' stop you, but *do* what you's
    gwine to do ;
For when you see a nigger foolin', den, sure as
    you are born,
You's gwine to see him comin' out de small end
    ob de horn.

I thanks you for de 'tention you hab gib dis after-
    noon ;
Sister Williams will oblige us by a raisin' ob a
    tune.
I see dat Brudder Johnson's gwine to pass around
    de hat ;
Don't let's hab no half-way doin's when it comes
    to dat.

Irwin Russell.

THE RAILROAD CROSSING.

I CAN'T tell much about the thing, 'twas done so
    powerful quick,
But 'pears to me I got a most outlandish heavy
    lick ;
It broke my leg, and tore my skulp, and jerked
    my arm most out.
But take a seat : I'll try and tell jest how it kem
    about.

You see, I'd started down to town, with that 'ere
    team of mine,
A-haulin' down a load o' corn to Ebenezer Kline

And drivin' slow; for, jest about a day or two
  before,
The off-horse run a splinter in his foot, and made
  it sore.

You know the railroad cuts across the road at
  Martin's Hole:
Well, thar I seed a great big sign, raised high
  upon a pole;
I thought I'd stop and read the thing, and find
  out what it said,
And so I stopped the hosses on the railroad-track,
  and read.

I ain't no scholar, rekollect, and so I had to
  spell:
I started kinder cautious like, with R-A-I- and L;
And that spelt "rail" as clear as mud; R-O-A-D
  was "road."
I lumped 'em: "railroad" was the word, and
  that 'ere much I knowed.

C-R-O and double S, with I-N-G to boot,
Made "crossing" jest as plain as Noah Webster
  dared to do't.
"Railroad crossing"—good enough!—L double-
  O-K, "look;"
And I was lookin' all the time, and spellin' like
  a book.

O-U-T spelt "out" jest right; and there it was,
  "look out:"
I's kinder cur'us, like to know jest what 'twas all
  about;

F-O-R and T-H-E; 'twas then "look out for
    the—"
And then I tried the next word; it commenced
    with E-N-G.

I'd got that fur, when suddintly there came an
    awful whack ;
A thousand fiery thunderbolts just scooped me
    off the track ;
The hosses went to Davy Jones, the wagon went
    to smash,
And I was h'isted seven yards above the tallest
    ash.

I didn't come to life ag'in fur 'bout a day or two ;
But, though I'm crippled up a heap, I sorter
    struggled through ;
It ain't the pain, nor 'tain't the loss o' that 'ere
    team of mine ;
But, stranger, how I'd like to know the rest of
    that ere sign !

                 Hezekiah Strong.

THE GRAND IMPOSITION HOTEL.

FROM the first minute I had give a thought to
goin' to see the Sentinal, my idee had been to git
boarded up in a private house. And I had my
eye (my mind's eye) upon who was willin' and
glad to board us. The Editor of the Auger'ses
wife's sister's husband's cousin boarded folks for
a livin' : she was a Dickey and married to a
Lampheare. The Editor of the Auger'ses wife

told me early in the spring that if she went she should go through the Sentinal to her sister's, and she happened to mention Miss Lampheare and the fact that she boarded up folks for a livin'. So when we decided to go, I told her when she wrote to her sister to ask her to ask Miss Lampheare if she was willin' to board Josiah and me, and how much she would ask for the boards. She wrote back ; her terms was moderate and inside of our means, and my mind was at rest. I almost knew that Josiah would want to throw himself onto his relatives through the Sentinal, but the underpinnin' was no firmer and rockier under our horse-barn than the determination of her that was Samantha Smith not to encamp upon a 2nd cousin. We had quite a lot of relations a-livin' out to Filadelfy,—though we never seen 'em,—sort o' distant, such as 2nd cousins, and so 4th, till they dwindled out o' bein' any relations at all ; descendants of the Daggets and Kidds,—Grandmother Allen was a Kidd,—no relation of old Captain Kidd. No! if any of his blood had been in my Josiah's veins, I would have bled him myself, if I had took a darnin'-needle to it. No! the Kidd'ses are likely folks, as I have heerd, and Josiah was rampant to go to cousin Sam Kidd's (a captain in the late war) through the Sentinal. But again I says to him, calmly and firmly,—

"No! Josiah Allen, no! anything but bringin' grief and trouble onto perfect strangers jest because they happened to be born second cousin to

you, unbeknown to 'em ;" and I repeated with
icy firmness,—for I see he was a-hankerin' awful-
ly,—"Josiah Allen, I will not encamp upon Captain Kidd through the Sentinal."

No ! Miss Lampheare was my theme and my
gole, and all boyed up with hope we arrove at
her dwellin'-place. Miss Lampheare met us at
the door herself. She was a tall spindlin'-look-
in' woman, one that had seen trouble,—for she
had always kep' boarders, and had had four husbands, and buried 'em in a row, her present one
bein' now in a decline. When I told her who I
was, she met me with warmth and said that any
friend of she that was Alminy Dickey was dear
to her.   But friendship, let it be ever so ardent,
cannot obtain cream from well-water, or cause
iron bedsteads to stretch out like Injy Rubber.
She had expected us sooner, if we come at all,
and her house was overflowin', every bed, lounge,
corner, and cupboard being occupied, and the
buro and stand drawers made up nightly for children.

What *was* we to do ? Night would soon let
down her cloudy mantilly upon Josiah and me,
and what was to become of us ? Miss Lampheare
seemed to pity us, and she directed us to a friend
of hers ; that friend was full ; he directed us to
another friend ; that friend was over-flowin'. And
so it went on till we was almost completely tired
out.

At last Josiah come out of a house, where he

had been seekin' rest and findin' it not; says
he,—

"They said mebbe we could git a room at the
'Grand Imposition Hotel.'" So we started off
there, Josiah a-scoldin' every step of the way, and
a-sayin',—

"I told you jest how it would be. We ort to
have gone to Captain Kidd's."

I didn't say nothin' back on the outside, for I
see by his face that it was no time for parley.
But my mind was firm on the inside, to board in
grocery-stores, and room under my umberell, before I threw myself onto a perfect stranger
through the Sentinal.

But a recital of our agony of mind will be of
little interest to the gay, and only sadden the
tender-hearted; and suffice it to say, in an hour's
time we was a-follerin' the hired man to a room
in the "Grand Imposition Hotel."

Our room was good enough, and big enough
for Josiah and me to turn round in one at a time.
It had a bed considerable narrer, but good and
healthy,—hard beds are considered healthy by
the best of doctors,—a chair, a lookin'-glass, and
a washstand. Josiah made a sight of fun of that,
because it didn't have but three legs.

But says I, firmly, "That is one more than
you have got, Josiah Allen." I wouldn't stand
none of his foolin'. . . .

When we eat supper we had a considerable
journey to the dinin'-room, which looked a good
deal on the plan of Miss Astor'ses, with lots of

colored folks a-goin' round, a-waitin' on the hungry crowd. I didn't see the woman of the house, —mebby she was laid up with a headache, or had gone out for an afternoon's visit,—but the colored waiters seemed to be real careful of her property ; they'd catch a teaspoon right out of their pocket and put it in your tea ; she couldn't have kep' a closer grip on her teaspoons herself.

I can truly say, without stretching the truth the width of a horsehair, that the chambermaid was as cross as a bear, for every identical thing I asked her for was a extra,—she couldn't do it without extra pay; but she did git me some ice-water once, without askin' me a cent extra for it. After we got to bed Josiah would lay and talk. He would speak out all of a sudden :

"Grand Imposition Hotel !"

And I'd say, " What of it ? What if it is ?"

And then he'd say, " They have got a crackin' good name, Samanthy. I love to see names that mean sunthin'." And then he'd ask me if I remembered the song about Barbara Allen, and if it would hurt my feelin's if he should lay and sing a verse of it to me, the bed put him in mind of it so.

I asked him what verse; but there was that in my tone that made him say no more about singin': he said it was the verse where Barbara wanted her mother to have her coffin made " long and narrer." And then he'd begin again about the pillars, and say how he wished he'd brought a couple of feathers from home to lay on, so he could have got some rest.

He had pulled out a little wad of cotton-battin' before we went to bed, to convince me of their ingredients.

But I says to him, "Josiah Allen, a easy conscience can rest even on cotton-battin' pillars," and I added, in awful meanin' tones, "*I* am sleepy, Josiah Allen, and waht to go to sleep. It is time," says I, with dignity, "that we was both reposin' in the arms of Morphine."

Nothin' quells him down quicker than to have me talk in a classical high-learnt way, and in a few minutes he was fast asleep.

[Mosquitoes and general misery kept Samantha awake till morning, when they sallied forth to do the Centennial.]

At last we reached the piazza, and emerged into the street. I see that every man, woman, and child was there in that identical street, and I thought to myself, there ain't no Sentinal to-day, and everybody has come out into this street for a walk. I knew it stood to reason that if there had been a Sentinal there would have been one or two men or wimmen attendin' to it, and I knew that every man, woman, and child on the hull face of the globe was right there before me, and behind me, and by the side of me, and fillin' the street full, walkin' afoot, and up in big covered wagons, all over 'em, on the inside, and hangin' on to the outside, as thick as bees a-swarmin'. Some of the horses was hitched ahead of each other, I s'pose so they could slip through the crowd easier. I couldn't see the village hardly any, owin' to the crowd a-crushin' of me ; but, from what little I *did* see, it was perfectly beautiful. I see they had fixed up for us : they had white-washed all their door-steps and winder-blinds white as snow, and trimmed the latter all off with black ribbon-strings.

Everything looked lovely and gay, and I thought, as I walked along, Jonesville couldn't compare with it for size and grandeur. I was a-walkin' along, crowded in body but happy in mind, when all of a sudden a thought come to me that goared me worse than any elbo or umberell that had pierced my ribs sense we sot out from the tavern. Thinks'es I all of a sudden, mebbe they have put off the Sentinal till I come ; mebbe I have disappointed the Nation, and belated 'em, and put 'em to trouble.

This was a sad thought, and wore on my mind considerable, and made me almost forget for the time bein' my bodily sufferin's as they pushed me this way and that, and goared me in the side with parasols and umberells, and carried off the tabs of my mantilly as far as they would go in every direction, and shoved, and stamped, and crowded. I declare, I was tore to pieces in mind and body, when I arrove at last at the entrance to the grounds. The crowd was fearful here, and the yells of different kinds was distractive : one conceited little creeter catched right holt of the tabs of my mantilly, and yelled right up in my face, " Won't you have a guide? Buy a guide,

mom, to the Sentinal." And seven or eight
others was a-yellin' the same thing to me, the
impudent creeters : I jest turned round and faced
the one that had got holt of my cape, and says
I,—

"Leggo of my tabs !"

"He wouldn't leggo ; he stood and yelled out
right up in my face, "Buy a guide : you hain't
got no guide !"

Says I, with dignity, "Yes, I have ; duty is my
guide, and also Josiah ; and now," says I firmly,
"if you don't leggo of my tabs, I'll *make* you
leggo." My mean skairt him ; he leggo, and I
follered on after my Josiah ; but where *was* Josiah ? I couldn't see him ; in tusslin' with that
impudent creeter over my cape, my companion
had got carried by the crowd out of my sight.
Oh, the agony of that half a moment ! I turned
and says to a policeman in almost agonizin'
tones,—

"Where is my Josiah ?"

He looked very polite at me, and says he,—

"I don't know."

Says I, "Find him for me instantly ! Have
you the heart to stand still and see husbands and
wives parted away from each other ? Have you
any principle about you ? Have you got entirely out of pity ?"

Says he, with the same polite look, "I don't
know."

"Have you a wife ?" says I, in thrilling axents.
"Have you any children ?"

I had heerd that there wasn't no information
to be extracted from them as a class, and I give
up; and I don't know what my next move would
have been if I hadn't catched sight of that beloved
face and that old familiar hat in front of me; I
hastened forred and kep' considerable calm in
mind, while my body was being crowded and
pushed round, for I thought if my conjectures
was true they would have reason enough to goar
me.

[Samantha and Josiah have no small trouble in learning the
art of paying their way into the Centennial, and the feelings
of the latter are outraged at being called an "adult" by the
gate-keeper. He doesn't relish being "called names" by a
ticket-seller. At length they get over the difficulty.]

We handed our fifty cents to a man, and he
dropped it down through a little slit in a counter;
and a gate that looked some like my new-fash-
ioned clothes-bars sort o' turned round with us
and let us in one at a time; and the minute I was
inside I see that my gloomy forebodin's had been
in vain : they hadn't put off the Sentinal for me!
That was my first glad thought; but my very
next thought was, Good land! and Good land!
and Good land! Them was my very first words,
and they didn't express my feelin's a half or even
a quarter. Why, comin' right out of that con-
tracted and crushin' crowd, it seemed as if the
place we found ourselves in was as roomy and
spacious as the desert of Sarah, s'posin' she, the
desert, was fixed off into a perfect garden of beauty, free for anybody to wander round and git
lost in.

And the majestic Main Building that nearly
loomed up in front of us! Why! if old Ocian
herself had turned into glass, and wood-work,
and cast iron, and shinin' ruffs, and towers, and
flags, and statues, and everything, and made a
glitterin' palace of herself, it couldn't (as it were)
have looked any more grand and imposin' and
roomy ; and if every sand by the sea-shore had
jumped up and put on a bunnet or hat, as the
case may be, there couldn't have been a bigger
crowd (seemingly) than there was passin' into it,
and a-passin' by, and a-paradin' round Josiah
and me.

Under these strange and almost apaulin' circumstances, is it any wonder that I stood stun
still, and said, out of the very depths of my heart,
the only words I could think of that would anywhere nigh express my feelin's, and they was
" Good land !"

But as my senses began to come back to me,
my next thought was, as I looked round on every
side of me, " Truly did my Josiah say that I
would see enough with one eye ;" and jest then a
band commenced playin' the " Star-Spangled
Banner." And hearing that soul-stirrin' music,
and seein' that very banner a-wavin' and floatin'
out, as if all the blue sky and rainbows sense
Noah's rainbow was cut up into its glorious
stripes, with the hull stars of heaven a-shinin' on
'em,—why, as my faculties come back to me,
a-seein' what I see, and heerin' what I heerd, I
thought of my 4 fathers, them 4 old fathers,
whose weak hands had first unfurled that banner

to the angry breeze, and thinks'es I, I would be willin' to change places with them 4 old men right here on the spot, to let them see in the bright sunshine of 1876 what they done in the cloudy darkness of 1776.

Marietta Holley.

FROM THE HONEYMOON.

Duke. You are welcome home.

Jul. Home! You are merry; this retired spot Would be a palace for an owl!

Duke. 'Tis ours.—

Jul. Ay, for the time we stay in it.

Duke. By Heaven,
This is the noble mansion that I spoke of!

Jul. This!—You are not in earnest, though you
 bear it
With such a sober brow. Come, come, you jest.

Duke. Indeed I jest not; were it ours in jest,
We should have none, wife.

Jul. Are you serious, sir?

Duke. I swear, as I'm your husband, and no
 duke.

Jul. No duke?

Duke. But of my own creation, lady.

Jul. Am I betrayed? Nay, do not play the fool!
It is too keen a joke.

Duke. You'll find it true.

Jul. You are no duke, then?

Duke. None.

Jul. Have I been cozened?
And have you no estate, sir?
No palaces, nor houses?

Duke. None but this :—

A small snug dwelling, and in good repair.

Jul. Nor money, nor effects?

Duke. None that I know of.

Jul. And the attendants who have waited on us?

Duke. They were my friends ; who, having
    done my business,

Are gone about their own.

Jul. Why, then, 'tis clear.—

That I was ever born !—What are you, sir ?

Duke. I am an honest man, that may content
    you.

Young, nor ill-favored, should not that content
you?

I am your husband, and that must content you.

Jul. I will go home !

Duke. You are at home already.

Jul. I'll not endure it !—But remember this :

Duke, or no duke, I'll be a duchess, sir !

Duke. A duchess ! You shall be a queen—
    to all

Who, by the courtesy, will call you so.

Jul. And I will have attendance !

Duke. So you shall,

When you have learned to wait upon yourself.

Jul. To wait upon myself ! Must I bear this ?

I could tear out my eyes, that bade you woo me,

And bite my tongue in two, for saying *yes !*

Duke. And if you should, 'twould grow again.

I think, to be an honest yeoman's wife

(For such, my would-be duchess, you will find me),

You were cut out by nature.

Jul. You will find, then,

That education, sir, has spoilt me for it.
Why! do you think I'll work?

    Duke. I think 'twill happen, wife.

    Jul. What! Rub and scrub
Your noble palace clean?

    ~~*Duke.* Those taper fingers~~
Will do it daintily.

    Jul. And dress your victuals
(If there be any)?—Oh! I could go mad!

    Duke. And mend my hose, and darn my night-
        caps neatly;
Wait, like an echo, till you're spoken to—

    Jul. Or like a clock, talk only once an hour?

    Duke. Or like a dial; for that quietly
Performs its work, and never speaks at all.

    Jul. To feed your poultry and your hogs!—
        Oh, monstrous!
And when I stir abroad, on great occasions
Carry a squeaking tithe pig to the vicar;
Or jolt with higglers' wives the market trot
To sell your eggs and butter!

    Duke. Excellent!
How well you sum the duties of a wife!
Why, what a blessing I shall have in you!

    Jul. A blessing!

    Duke. When they talk of you and me,
Darby and Joan shall no more be remembered:—
We shall be happy!

    Jul. Shall we?

    Duke. Wondrous happy!
Oh, you will make an admirable wife!

    Jul. I will make a vixen.

    Duke. What?

Jul. A very vixen.

Duke. Oh, no! We'll have no vixens.

Jul. I'll not bear it!
I'll to my father's—

Duke. Gently: you forget
You are a perfect stranger to the road.

Jul. My wrongs will find a way, or make one.

Duke. Softly!
You stir not hence, except to take the air;
And then I'll breathe it with you.

Jul. What, confine me?

Duke. 'Twould be unsafe to trust you yet
    abroad.

Jul. Am I a truant schoolboy?

Duke. Nay, not so;
But you must keep your bounds.

Jul. And if I break them
Perhaps you'll beat me?

Duke. Beat you!
The man that lays his hand upon a woman,
Save in the way of kindness, is a wretch
Whom 'twere gross flattery to name a coward.—
I'll talk to you, lady, but not beat you.

Jul. Well, if I may not travel to my father,
I may write to him, surely!—And I will—
If I can meet within your spacious dukedom
Three such unhoped-for miracles at once,
As pens, and ink, and paper.

Duke. You will find them
In the next room.—A word, before you go:
You are my wife, by every tie that's sacred;
The partner of my fortune—

Jul. Your fortune!

Duke. Peace !—No fooling, idle woman !
Beneath th' attesting eye of Heaven I've sworn
To love, to honor, cherish, and protect you.
No human power can part us. What remains,
    then ?
To fret, and worry, and torment each other,
And give a keener edge to our hard fate
By sharp upbraidings and perpetual jars ?
Or, like a loving and a patient pair
(Waked from a dream of grandeur, to depend
Upon their daily labor for support),
To soothe the taste of fortune's lowliness
With sweet consent and mutual fond endear-
    ment ?
Now to your chamber—write whate'er you please;
But pause before you stain the spotless paper
With words that may inflame, but cannot heal !
    Jul. Why, what a patient worm you take me
    for !
    Duke. I took you for a wife ; and ere I've done,
I'll know you for a good one.
    Jul. You shall know me
For a right woman, full of her own sex ;
Who, when she suffers wrong, will speak her
    anger ;
Who feels her own prerogative, and scorns,
By the proud reason of superior man,
To be taught patience, when her swelling heart
Cries out revenge !   [*Exit.*
    Duke. Why, let the flood rage on !
There is no tide in woman's wildest passion
But hath an ebb.—I've broke the ice, however.—
Write to her father !—She may write a folio.—

But if she send it!—'Twill divert her spleen:
The flow of ink may save her blood-letting.
Perchance she may have fits!—They are seldom
 mortal,
Save when the Doctor's sent for.
Though I have heard some husbands say, and
 wisely,
A woman's honor is her safest guard,
Yet there's some virtue in a lock and key.
So, thus begins our honeymoon.—'Tis well!
For the first fortnight, ruder than March winds,
She'll blow a hurricane.   The next, perhaps,
Like April she may wear a changeful face
Of storm and sunshine: and when that is past,
She will break glorious as unclouded May;
And where the thorns grew bare, the spreading
 blossoms
Meet with no lagging frost to kill their sweet-
 ness.
Whilst others, for a month's delirious joy,
Buy a dull age of penance, we, more wisely,
Taste first the wholesome bitter of the cup,
That after to the very lees shall relish;
And to the close of this frail life prolong
The pure delights of a well-governed marriage.
        John Tobin.

JOHN GILPIN'S RIDE.

JOHN GILPIN was a citizen of credit and renown;
A train-band captain eke was he, of famous London town.
John Gilpin's spouse said to her dear, "Though
 wedded we have been

These twice ten tedious years, yet we no holiday
    have seen.

"To-morrow is our wedding-day, and we shall
    then repair
Unto the Bell at Edmonton, all in a chaise-and-
    pair.
My sister and my sister's child, myself and chil-
    dren three,
Will fill the chaise : so you must ride on horse-
    back after we."

He soon replied, "I do admire, of womankind,
    but one,
And you are she, my dearest dear, therefore it
    shall be done.
I am a linen-draper bold, as all the world doth
    know,
And my good friend the calender will lend his
    horse to go."

Quoth Mrs. Gilpin, "That's well said; and, for
    that wine is dear,
We will be furnished with our own, which is both
    bright and clear."
John Gilpin kissed his loving wife : o'erjoyed was
    he to find
That, though on pleasure she was bent, she had
    a frugal mind.

The morning came; the chaise was brought, but
    yet was not allowed
To drive up to the door, lest all should say that
    she was proud.

So three doors off the chaise was stayed, where
    they did all get in,—
Six precious souls, and all agog to dash through
    thick and thin!

Smack went the whip, round went the wheels;
    were never folks so glad;
The stones did rattle underneath, as if Cheapside
    were mad.
John Gilpin, at his horse's side, seized fast the
    flowing mane,
And up he got, in haste to ride, but soon came
    down again;

For saddle-tree scarce reached had he, his jour-
    ney to begin,
When, turning round his head, he saw three cus-
    tomers come in.
So down he came; for loss of time, although it
    grieved him sore,
Yet loss of pence, full well he knew, would trouble
    him much more.

'Twas long before the customers were suited to
    their mind,
When Betty screaming came down-stairs, "The
    wine is left behind!"
"Good lack!" quoth he; "yet bring it me, my
    leathern belt likewise,
In which I wear my trusty sword, when I do ex-
    ercise."

Now, Mrs. Gilpin (careful soul!) had two stone
    bottles found,
To hold the liquor that she loved, and keep it
    safe and sound;

Each bottle had a curling ear, through which the
    belt he drew,
And hung a bottle on each side, to make his bal-
    ance true.

Then over all, that he might be equipped from
    top to toe,
His long red cloak, well brushed and neat, he
    manfully did throw.
Now see him mounted once again upon his nim-
    ble steed,
Full slowly pacing o'er the stones with caution
    and good heed.

But finding soon a smoother road beneath his
    well-shod feet,
The snorting beast began to trot, which galled
    him in his seat.
"So! fair and softly!" John he cried; but John
    he cried in vain;
The trot became a gallop soon, in spite of curb
    and rein.

So, stooping down, as needs he must who cannot
    sit upright,
He grasped the mane with both his hands, and
    eke with all his might.
His horse, who never in that sort had handled
    been before,
What thing upon his back had got did wonder
    more and more.

Away went Gilpin, neck or naught; away went
    hat and wig;

He little dreamed, when he set out, of running
    such a rig.
The wind did blow, the cloak did fly, like stream-
    er long and gay,
Till, loop and button failing both, at last it flew
    away.

Then might all people well discern the bottles he
    had slung;
A bottle swinging at each side, as hath been said
    or sung.
The dogs did bark, the children screamed, up
    flew the windows all,
And every soul cried out, " Well done !" as loud
    as he could bawl.

Away went Gilpin, who but he ! his fame soon
    spread around,
" He carries weight ! He rides a race ! 'Tis for
    a thousand pound !"
And still, as fast as he drew near, 'twas wonder-
    ful to view
How in a trice the turnpike-men their gates wide
    open threw.

And now, as he went bowing down his reeking
    head full low,
The bottles twain, behind his back, were shat-
    tered at a blow.
Down ran the wine into the road, most piteous
    to be seen,
Which made his horse's flanks to smoke, as they
    had basted been.

But still he seemed to carry weight, with leather
    girdle braced,
For all might see the bottle-necks still dangling
    at his waist.
Thus all through merry Islington these gambols
    he did play,
And till he came unto the Wash of Edmonton so
    gay.

And there he threw the Wash about on both
    sides of the way,
Just like unto a trundling-mop, or a wild goose
    at play.
At Edmonton his loving wife, from the balcony,
    spied
Her tender husband, wondering much to see how
    he did ride.

"Stop, stop, John Gilpin! here's the house!"
    they all aloud did cry;
The dinner waits, and we are tired!" Said Gil-
    pin, "So am I!"
But yet his horse was not a whit inclined to tarry
    there;
For why? his owner had a house full ten miles
    off, at Ware.

So like an arrow swift he flew shot by an archer
    strong,
So did he fly,—which brings me to the middle of
    my song.
Away went Gilpin, out of breath, and sore against
    his will,

Till at his friend the calender's his horse at last
    stood still.

The calender, amazed to see his friend in such a
    trim,
Laid down his pipe, flew to the gate, and thus
    accosted him:
"What news? What news? Your tidings tell!
    Tell me you must and shall!
Say why bare-headed you are come,—or why you
    come at all."

Now, Gilpin had a pleasant wit, and loved a
    timely joke,
And thus unto the calender in merry guise he
    spoke:
"I came because your horse would come; and,
    if I well forbode,
My hat and wig will soon be here; they are upon
    the road!"

The calender, right glad to find his friend in
    merry pin,
Returned him not a single word, but to the house
    went in,
Whence straight he came with hat and wig,—a
    wig that flowed behind,
A hat not much the worse for wear,—each comely
    in its kind.

He held them up, and in his turn thus showed
    his ready wit:
"My head is twice as big as yours: they there-
    fore needs must fit.

354 CORRECT AND EFFECTIVE ELOCUTION.

But let me scrape the dirt away that hangs upon
    your face ;
And stop and eat, for well you may be in a hun-
    gry case."

Said John, "It is my wedding-day, and all the
    world would stare
If wife should dine at Edmonton and I should
    dine at Ware."
So, turning to his horse, he said, "I am in haste
    to dine ;
Twas for your pleasure you came here, you shall
    go back for mine."

Ah, luckless speech and bootless boast! for which
    he paid full dear ;
For, while he spake, a braying ass did sing most
    loud and clear,
Whereat his horse did snort as he had heard a
    lion roar,
And galloped off with all his might, as he had
    done before.

Away went Gilpin, and away went Gilpin's hat
    and wig :
He lost them sooner than at first; for why?—
    they were too big.
Now, Mistress Gilpin, when she saw her husband
    posting down
Into the country far away, she pulled out half a
    crown,

And thus unto the youth she said that drove
    them to the Bell,

"This shall be yours when you bring back my
    husband safe and well."
The youth did ride, and soon did meet John
    coming back amain,
Whom in a trice he tried to stop, by catching at
    his rein ;

But, not performing what he meant, and gladly
    would have done,
The frighted steed he frighted more, and made
    him faster run.
Away went Gilpin, and away went post-boy at
    his heels,
The post-boy's horse right glad to miss the lum-
    bering of the wheels.

Six gentlemen upon the road, thus seeing Gilpin
    fly,
With post-boy scampering in the rear, they raised
    the hue and cry :
"Stop thief ! Stop thief !—a highwayman !"—
    not one of them was mute,
And all and each that passed that way did join
    in the pursuit.

And now the turnpike gates again flew open in
    short space,
The toll-men thinking, as before, that Gilpin rode
    a race.
And so he did, and won it too, for he got first to
    town,
Nor stopped till where he had got up he did
    again get down.

Now let us sing, " long live the king," and Gil-
    pin, long live he,
And when he next doth ride abroad may I be
    there to see.

William Cowper.

A PROFITABLE SHOT.

TOM SHERIDAN used to tell a story *for* and
against himself, which we shall take leave to relate.

He was staying at Lord Craven's, at Benham
(or rather Hempstead), and one day proceeded
on a shooting-excursion, like Hawthorn, with
only " his dog and his gun," on foot, and unat-
tended by companion or keeper. The sport was
bad, the birds few and shy, and he walked and
walked in search of game, until unconsciously he
entered the domain of some neighboring squire.
A very short time after he perceived advancing
towards him, at the top of his speed, a jolly, com-
fortable-looking gentleman, followed by a servant, armed, as it appeared, for conflict. Tom
took up a position, and waited the approach of
the enemy.

"Halloo! you, sir," said the squire, when
within half ear-shot, " what are you doing here,
sir, eh?"

" I'm shooting, sir," said Tom.

" Do you know where you are, sir?" said the
squire.

" I'm here, sir," said Tom.

"Here, sir!" said the squire, growing angry; "and do you know where here *is*, sir? These, sir, are *my* manors: what d'ye think of that, sir?"

"Why, sir, as to your manners," said Tom, "I can't say they seem over-agreeable."

"I don't want any jokes, sir," said the squire: "I hate jokes. Who are you, sir?—what are you?"

"Why, sir," said Tom, "my name is Sheridan; I am staying at Lord Craven's; I have come out for some sport; I have not had any, and am not aware that I am trespassing."

"Sheridan!" said the squire, cooling a little; "oh, from Lord Craven's, eh? Well, sir, I could not know that, sir, I—"

"No, sir," said Tom, "but you need not have been in a passion."

"Not in a passion, Mr. Sheridan!" said the squire; "you don't know what these preserves have cost me, and the pains and trouble I have been at with them. It's all very well for you to talk, but if you were in my place I would like to know what *you* would say upon such an occasion."

"Why, sir," said Tom, "if I were in *your* place, under all the circumstances, I should say, 'I am convinced, Mr. Sheridan, you did not mean to annoy me; and as you look a good deal tired, perhaps you will come up to my house and take some refreshment.'"

The squire was hit hard by this nonchalance,

and (as the newspapers say), " it is needless to add," acted upon Sheridan's suggestion.

" So far," said poor Tom, " the story tells for me. Now you shall hear the sequel."

After having regaled himself at the squire's house, and having said five hundred more good things than he swallowed, having delighted his host, and more than half won the hearts of his wife and daughter, the sportsman proceeded on his return homewards.

In the course of his walk he passed through a farm-yard : in the front of the farm-house was a green, in the centre of which was a pond, in the pond were ducks innumerable, swimming and diving ; on its verdant banks a motley group of gallant cocks and pert partlets, picking and feed-ing : the farmer was leaning over the thatch of his barn, which stood near two cottages on the side of the green.

Tom hated to go back with an empty bag ; and, having failed in his attempts at higher game, it struck him as a good joke to ridicule the ex-ploits of the day himself, in order to prevent any one else from doing it for him ; and he thought that to carry home a certain number of the do-mestic inhabitants of the pond and its vicinity would serve the purpose admirably.   Accord-ingly, up he goes to the farmer, and accosts him very civilly :

" My good friend," says Tom, " I'll make you an offer."

" Of what, sir ?" says the farmer.

" Why," replies Tom, " I have been out all day fagging after birds, and haven't had a shot ; now, both my barrels are loaded, I should like to take home something : what shall I give you to let me have a shot with each barrel at those ducks and fowls,—I standing here, and to have whatever I kill ?"

" What sort of a shot are you ?" said the farmer.

" Fairish," said Tom ; " fairish."

" And to have all you kill ?" said the farmer, " eh ?"

" Exactly so," said Tom.

" Half a guinea," said the farmer.

" That's too much," said Tom. " I'll tell you what I'll do : I'll give you a seven-shilling piece, which happens to be all the money I have in my pocket."

" Well," said the man, " hand't over."

The payment was made ; Tom, true to his bargain, took his post by the barn door, and let fly with one barrel, and then with the other ; and such quacking and splashing and screaming and fluttering had never been seen in that place before.

Away ran Tom, and, delighted at his success, picked up first a hen, then a chicken, then fished out a dying duck or two, and so on, until he numbered eight head of domestic game, with which his bag was nobly distended.

" Those were right good shots, sir," said the farmer.

"Yes," said Tom; "eight ducks and fowls are more than you bargained for, old fellow,—worth rather more, I suspect, than seven shillings,—eh?"

"Why, yes," said the man, scratching his head, "I think they be; but what do I care for that? *they are none of mine!*"

"Here," said Tom, "I was for once in my life beaten, and made off as fast as I could, for fear the right owner of the game might make his appearance; not but that I could have given the fellow that took me in seven times as much as I did, for his cunning and coolness."

George A. Sala.

LADY CLARE.

IT was the time when lilies blow,
   And the clouds are highest up in air,
Lord Ronald brought a lily-white doe
   To give his cousin, Lady Clare.

I trow they did not part in scorn;
   Lovers long betrothed were they:
They two will wed the morrow morn;
   God's blessing on the day!

"He does not love me for my birth,
   Nor for my lands as broad and fair;
He loves me for my own true worth,
   And that is well," said Lady Clare.

In there came old Alice the nurse,
  Said, " Who was this that went from thee?"
"It was my cousin," said Lady Clare,
  "To-morrow he weds with me."

"O God be thanked!" said Alice the nurse,
  " That all comes round so just and fair,
Lord Ronald is heir of all your lands,
  And you are not the Lady Clare."

" Are ye out of your mind, my nurse, my nurse?"
  Said Lady Clare, " that ye speak so wild?"
"As God's above," said Alice the nurse,
  " I speak the truth; you are my child."

"The old earl's daughter died at my breast;
  I speak the truth as I live by bread;
I buried her like my own sweet child,
  And put my child in her stead."

"Falsely, falsely have ye done,
  O mother," she said, " if this be true,
To keep the best man under the sun
  So many years from his due."

" Nay now, my child," said Alice the nurse,
  "But keep the secret for your life,
And all you have will be Lord Ronald's,
  When you are man and wife."

" If I'm a beggar born," she said,
  "I will speak out, for I dare not lie.
Pull off, pull off the broach of gold,
  And fling the diamond necklace by."

" Nay now, my child," said Alice the nurse,
  " But keep the secret all ye can."
She said " Not so : but I will know
  If there be any faith in man."

" Nay now, what faith ?" said Alice the nurse,
  " The man will cleave unto his right."
" And he shall have it," the lady replied,
  " Tho' I should die to-night."

" Yet give one kiss to your mother dear!
  Alas, my child, I sinned for thee."
" O mother, mother, mother," she said,
  " So strange it seems to me.

" Yet here's a kiss for my mother dear,
  My mother dear, if this be so,
And lay your hand upon my head,
  And bless me, mother, ere I go."

She clad herself in a russet gown,
  She was no longer Lady Clare,
She went by dale and she went by down,
  With a single rose in her hair.

The lily-white doe Lord Ronald had brought
  Leapt up from where she lay,
Dropt her head in the maiden's hand
  And followed her all the way.

Down stept Lord Ronald from his tower,
  " Lady Clare, you shame your worth,
Why come you drest like a village maid,
  That are the flower of the earth ?"

SELECTIONS FOR READING AND RECITATION. 363

"If I come drest like a village maid,
    I am but as my fortunes are;
I am a beggar born," she said,
    "And not the Lady Clare."

"Play me no tricks," said Lord Ronald,
    "For I am yours in word and deed;
Play me no tricks," said Lord Ronald,
    "Your riddle is hard to read."

Oh, and proudly stood she up;
    Her heart within her did not fail:
She looked into Lord Ronald's eyes
    And told him all her nurse's tale.

He laughed a laugh of merry scorn,
    He turned and kissed her where she stood.
"If you are not the heiress born,
    And I," said he, "the next of blood,"

"If you are not the heiress born,
    And I," said he, "the lawful heir,
We two will wed to-morrow morn,
    And you shall still be Lady Clare."

                                        Tennyson.

THE TAMING OF BUCEPHALUS.

"BRING forth the steed!" It was a level plain,
Broad and unbroken as the mighty sea
When in their prison-caves the winds lie
        chained.
There Philip sat, pavilioned from the sun;

There, all around, thronged Macedonia's hosts,
Bannered, and plumed, and armed—a vast array!
Then Philip waved his sceptre. Silence fell
O'er all the plain. 'Twas but a moment's pause;
"Obey my son, Pharsalian! bring the steed!"
The monarch spoke. A signal to the grooms,
And on the plain they led *Bucephalus.*
"Mount, vassal, mount! Why pales thy cheek
    with fear?
"Mount!—ha! art slain? Another: mount
    again!"
'Twas all in vain. No hand could curb a neck,
Clothed with such might and grandeur, to the
    rein.
No thong or spur could make his fury yield.
Now bounds he from the earth; and now he
    rears—
Now madly plunges—strives to rush away,
Like that strong bird, his fellow-king of air!

              Then Alexander threw
His light cloak from his shoulders, and drew
    nigh.
The brave steed was no courtier; prince and
    groom
Bore the same mien to him. He started back;
But with firm grasp the youth retained, and
    turned
His fierce eyes from his shadow to the sun.
Then, with that hand, in after times which hurled
The bolts of war among embattled hosts,
Conquered all Greece, and over Persia swayed

Imperial command—which on Fame's temple
Graved, ALEXANDER, VICTOR OF THE WORLD!—
With that bold hand he smoothed the flowing
    mane,
Patted the glossy skin with soft caress,
Soothingly speaking in low voice the while,
Lightly he vaulted to his first great strife.
How like a Centaur[1] looked the steed and youth!
Firmly the hero sat; his glowing cheek
Flushed with the rare excitement: his high brow
Pale with a stern resolve: his lip as smiling,
And his glance as calm, as if, in dalliance,
Instead of danger, with a girl he played.
Untutored to obey, how raves the steed!
Champing the bit, and tossing the white foam,
And struggling to be free, that he might dart,
Swift as an arrow from a shivering bow.
The rein is loosened. "Now, Bucephalus!"
Away! away!—he flies, away—away!
The multitude stood hushed, in breathless awe,
And gazed into the distance.

                Lo! a speck—
A darksome speck, on the horizon! 'Tis—
'Tis he! Now it enlarges; now are seen
The horse and rider; now, with ordered pace,
The horse approaches, and the rider leaps

---

[1] The first men who tamed horses and rode them were supposed to be part of the horse, and were called *Centaurs*. Prescott, in his History of the Conquest of Mexico, says that the Mexicans, who had never seen a horse before, made the same mistake in regard to the cavalry of the Spanish invaders.

Down to the earth, and bends his rapid pace
Unto the king's pavilion.   The wild steed,
Unled, uncalled, is following his subduer.
Philip wept tears of joy: "My son, go seek
A larger empire; for so vast a soul
Too small is Macedonia!"

Park Benjamin.

ANNE HATHAWAY.

ONCE on a time, when jewels flashed,
And moonlit fountains softly splashed,
And all the air was sweet and bright
With music, mirth, and deft delight,
A courtly dame drew, laughing, near
  A poet—greatest of his time,
And chirped a question in his ear,
    With voice like silver bells in chime:
" Good Mr. Shakespeare, I would know
    What name thy lady bore, in sooth,
Ere thine.   Nay, little time ago
    It seems—for we still mark her youth;
Some high-born name, I trow, and yet,
Although I've heard it, I forget."
Then answered he with dignity,
Yet blithely, for the hour was gay,
" My Lady's name?   Anne Hathaway."

" And good, sweet sir," the dame pursued,
Too fair and winsome to be rude,
" 'Tis whispered here, and whispered there,
By doughty knights and ladies fair,

That—that—well, that her loyal lord
  Doth e'en obey her slightest will.
Now, my good lord—I pledge my word—
  Though loving well, doth heed me ill;
How art thou conquered, prithee, tell,"
  She pleaded with her pretty frown;
"I fain would know what mighty spell
  Can bring a haughty husband down."
She ceased, and raised her eager face
To his, with laughing, plaintive grace.
Then answered he with dignity,
Yet blithely, for the hour was gay,
"Ah, Lady! I can only say
Her name again—*Anne Hath-a-way.*"

THE INQUIRY.

Tell me, ye winged winds, that round my pathway roar,
Do ye not know some spot where mortals weep
  no more?
Some lone and pleasant dell, some valley in the
  west,
Where, free from toil and pain, the weary soul
  may rest?
    The loud wind dwindled to a whisper low,
    And sighed for pity as it answered—"No."

Tell me, thou mighty deep, whose billows round
  me play,
Know'st thou some favored spot, some island far
  away,

Where weary man may find the bliss for which
    he sighs—
Where sorrow never lives, and friendship never
    dies ?
    The loud waves rolling in perpetual flow
    Stopped for a while, and sighed to answer—
      " No."

And thou, serenest moon, that, with such lovely
    face,
Dost look upon the earth, asleep in night's em-
    brace,
Tell me, in all thy round, hast thou not seen
    some spot
Where miserable man might find a happier lot?
    Behind a cloud the moon withdrew in woe,
    And a voice, sweet, but sad, responded—
      " No."

Tell me, my secret soul—oh, tell me, Hope and
    Faith,
Is there no resting-place from sorrow, sin and
    death ?—
Is there no happy spot where mortals may be
    blessed,
Where grief may find a balm, and weariness a
    rest ?
    Faith, Hope, and Love, best boons to mortals
      given,
    Waved their bright wings, and whispered—
      " YES, IN HEAVEN !"

                  Charles Mackay.

THE LAUNCHING OF THE SHIP.

ALL is finished! and at length
Has come the bridal day
Of beauty and of strength.
To-day the vessel shall be launched!
With fleecy clouds the sky is blanched,
And o'er the bay,
Slowly, in all his splendors dight,
The great sun rises to behold the sight.

The ocean old,
Centuries old,
Strong as youth, and as uncontrolled,
Paces restless to and fro,
Up and down the sands of gold.
His beating heart is not at rest;
And far and wide,
With ceaseless flow,
His beard of snow
Heaves with the heaving of his breast.

He waits impatient for his bride.
There she stands,
With her foot upon the sands,
Decked with flags and streamers gay,
In honor of her marriage day,
Her snow-white signals fluttering, blending,
Round her like a veil descending,
Ready to be
The bride of the gray, old sea.

Then the Master,
With a gesture of command,
Waved his hand;
And at the word,
Loud and sudden there was heard,
All around them and below,
The sound of hammers, blow on blow,
Knocking away the shores and spurs.

And see! she stirs!
She starts,—she moves,—she seems to feel
The thrill of life along her keel,
And, spurning with her foot the ground,
With one exulting, joyous bound,
She leaps into the ocean's arms!

And lo! from the assembled crowd
There rose a shout, prolonged and loud,
That to the ocean seemed to say,
"Take her, O bridegroom, old and gray,
Take her to thy protecting arms,
With all her youth and all her charms!"

How beautiful she is! how fair
She lies within those arms, that press
Her form with many a soft caress
Of tenderness and watchful care!
Sail forth into the sea, O ship!
Through wind and wave, right onward steer!
The moistened eye, the trembling lip,
Are not the signs of doubt or fear.

Thou, too, sail on, O ship of State!
Sail on, O Union, strong and great!
Humanity, with all its fears,
With all the hopes of future years,
Is hanging breathless on thy fate!
We know what Master laid thy keel,
What workmen wrought thy ribs of steel,
Who made each mast, and sail, and rope,
What anvils rang, what hammers beat,
In what a forge, and what a heat,
Were shaped the anchors of thy hope!

Fear not each sudden sound and shock;
'Tis of the wave, and not the rock;
'Tis but the flapping of the sail,
And not a rent made by the gale.
In spite of rock and tempest roar,
In spite of false lights on the shore,
Sail on, nor fear to breast the sea!
Our hearts, our hopes, are all with thee,
Our hearts, our hopes, our prayers, our tears,
Our faith triumphant o'er our fears,
Are all with thee—are all with thee!

Longfellow.

THE HIGH TIDE ON THE COAST OF LINCOLNSHIRE.

(1571.)

The old mayor climbed the belfry tower,
  The ringers ran by two, by three;
" Pull, if ye never pulled before;
  Good ringers, pull your best," quoth he,

372 CORRECT AND EFFECTIVE ELOCUTION.

"Play uppe, play uppe, O Boston bells !
Ply all your changes, all your swells,
  Play uppe 'The Brides of Enderby.'"

Men say it was a stolen tyde—
  The Lord that sent it, He knows all ;
But in myne ears doth still abide
  The message that the bells let fall :
And there was naught of strange, beside
The flight of mews and peewits pied
  By millions crouched on the old sea wall.

I sat and spun within the doore,
  My thread break off, I raised myne eyes ;
The level sun, like ruddy ore,
  Lay sinking in the barren skies ;
And dark against day's golden death
She moved where Lindis wandereth,
My sonne's faire wife, Elizabeth.

"Cusha ! Cusha ! Cusha !" calling,
Ere the early dews were falling,
Farre away I heard her song,
"Cusha ! Cusha !" all along ;
Where the reedy Lindis floweth,
  Floweth, floweth,
From the meads where melick groweth
Faintly came her milking song.*

   * "An important section of Scandinavian songs are the herds-
men's. Their age it is impossible to state, but they all bear the
same character. The herdsman or maiden calls home the
cattle from the mountain-side, either with the cow-horn or
Lur, or by singing a melody, with the *echo* formed on the
intervals of that instrument."—GROVE.

Andante.
p
Cu - sha! cu - sha! cu - sha! cu - sha!
For the dews will soone be fal - ling;
Leave your mea - dow grass - es mel - low,
sf
dim.
Quit your cow - slips, cow - slips yel - low.
cres.
Come uppe, White - foot, come uppe, Light - foot,
Come uppe, White - foot, come uppe, Light - foot,
cres.
Quit the stalks of pars - ley hol - low;
Quit the stalks of pars - ley hol - low;
sf
Come uppe, Jet - ty, rise and fol - low,
Come uppe, Jet - ty, rise and fol - low,

If it be long, ay, long ago,
  When I beginne to think howe long,
Againe I hear the Lindis flow,
  Swift as an arrowe, sharpe and strong;
And all the aire, it seemeth mee,
Bin full of floating bells (sayth shee),
That ring the tune of Enderby.

Alle fresh the level pasture lay,
  And not a shadowe mote be seene,
Save where full fyve good miles away
  The steeple towered from out the greene;
And lo! the great bell farre and wide
Was heard in all the country side
That Saturday at eventide.

The swanherds where their sedges are
  Moved on in sunset's golden breath,
The shepherde lads I heard afarre,
  And my sonne's wife, Elizabeth;
Till floating o'er the grassy sea
Came downe that kyndly message free,
The " Brides of Mavis Enderby."

Then some looked uppe into the sky,
  And all along where Lindis flows
To where the goodly vessels lie,
  And where the lordly steeple shows.
They sayde, " And why should this thing be?
What danger lowers by land or sea?
They ring the tune of Enderby!

" For evil news from Mablethorpe,
  Of pyrate galleys warping down;

For shippes ashore beyond the scorpe,
   They have not spared to wake the towne ;
But while the west bin red to see,
And storms be none, and pyrates flee,
Why ring ' The Brides of Enderby ? ' "

I looked without, and lo ! my sonne
   Came riding downe with might and main :
He raised a shout as he drew on,
   Till all the welkin rang again,
" Elizabeth ! Elizabeth !"
(A sweeter woman ne'er drew breath
Than my sonne's wife Elizabeth.)

" The olde sea wall (he cried) is downe,
   The rising tide comes on apace,
And boats adrift in yonder towne
   Go sailing uppe the market-place."
He shook as one that looks on death :
" God save you, mother !" straight he saith ;
" Where is my wife, Elizabeth ?"

" Good sonne, where Lindis winds her way,
   With her two bairns I marked her long ;
And ere young bells beganne to play
   Afar I heard her milking song."
He looked across the grassy lea,
To right, to left, " Ho, Enderby !"
They rang " The Brides of Enderby !"

With that he cried and beat his breast ;
   For lo ! along the river's bed
A mighty eygre reared his crest,
   And uppe the Lindis raging sped.

It swept with thunderous noises loud;
Shaped like a curling snow-white cloud,
Or like a demon in a shroud.

And rearing Lindis backward pressed
　　Shook all her trembling bankes amaine;
Then madly at the eygre's breast
　　Flung uppe her weltering walls again
Then bankes came downe with ruin and rout—
Then beaten foam flew round about—
Then all the mighty floods were out.

So farre, so fast the eygre drave,
　　The heart had hardly time to beat
Before a shallow seething wave
　　Sobbed in the grasses at our feet:
The feet had hardly time to flee
Before it brake against the knee,
And all the world was in the sea.

Upon the roofe we sate that night,
　　The noise of bells went sweeping by;
I marked the lofty beacon light
　　Stream from the church tower, red and high—
A lurid mark and dread to see;
And awesome bells they were to mee,
That in the dark rang "Enderby."

They rang the sailor lads to guide
　　From roofe to roofe who fearless rowed;
And I—my sonne was at my side,
　　And yet the ruddy beacon glowed;

And yet he moaned beneath his breath,
"O come in life, or come in death !
O lost ! my love, Elizabeth."

And didst thou visit him no more?
   Thou didst, thou didst, my daughter deare ;
The waters laid thee at his doore,
   Ere yet the early dawn was clear.
Thy pretty bairns in fast embrace,
The lifted sun shone on thy face,
Downe drifted to thy dwelling-place.

That flow strewed wrecks about the grass,
   That ebbe swept out the flocks to sea ;
A fatal ebbe and flow, alas !
   To manye more than myne and mee :
But each will mourn his own (she saith) ;
And sweeter woman ne'er drew breath
Than my sonne's wife, Elizabeth.

I shall never hear her more
By the reedy Lindis shore,
"Cusha ! Cusha ! Cusha !" calling,
Ere the early dews be falling ;
I shall never hear her song,
"Cusha ! Cusha !" all along
Where the sunny Lindis floweth,
    Goeth, floweth ;
From the meads where melick groweth,
When the water, winding down,
Onward floweth to the town.

I shall never see her more
Where the reeds and rushes quiver,
    Shiver, quiver;
Stand beside the sobbing river
Sobbing, throbbing, in its falling
To the sandy lonesome shore;
I shall never hear her calling,
"Leave your meadow grasses mellow,
    Mellow, mellow;
Quit your cowslips, cowslips yellow;
Come uppe, Whitefoot, come uppe, Lightfoot;
Quit your pipes of parsley hollow,
    Hollow, hollow;
Come uppe, Lightfoot, rise and follow;
  Lightfoot, Whitefoot,
From your clovers lift the head:
Come uppe, Jetty, follow, follow,
Jetty, to the milking shed."—*Jean Ingelow.*

THE BELL OF ZANORA.

THE ruddy sun was setting behind the Murchian
    hills,
The fields were warmed to splendor and golden
    flowed the rills.
Across the little valley, where lay the Spanish
    town,
The dying sun's last blessing, a glory, floated
    down.
Amid the fields the peasants led in the grazing
    kine,
And faintly came a tinkling as trudged the peace-
    ful line.

Upon the height the convent, a ruin old and gray,
Towered upward, and its shadow across the val-
    ley lay.
Before that ancient ruin, prone on the scented
    grass,
A boy of fifteen summers watched day's bright
    glory pass :
The lad was there on duty and oft about him
    scanned.
Zanora feared the coming of robber Gomez's
    band ;
Of Gomez, fierce and heartless, the terror of the
    vale,
Whose name made women shudder and bravest
    . men grow pale.
Unto the town a rumor that Gomez fierce would
    come
And sack the peaceful hamlet made stoutest
    hearts all dumb.
The peasants cleaned their firelocks, the women
    watched and prayed
That the band of robber Gomez upon its path be
    stayed.
Yet time wore on, and scathless still stood the
    little town,
But from its ancient convent a watcher still looked
    down.
For clear from 'neath its portals each roadway
    might be scanned,
And there from morn till night they watched for
    Gomez's band.

The old bell of the convent within its tower still
 hung,
Its ropes with dangling curves seemed waiting
 to be rung,
For if a sight of Gomez came to the watcher there,
He straight would let the old bell with warning
 fill the air,
Unto the town a signal to rally fast and stand,
And, ready for the onslaught, beat back the rob-
 ber band.
This day was Rooe watcher until the sun hung
 low,
And then, with watching wearied, he lay and
 gazed below.
He watched the smoke that floated above his
 mother's cot.
To him the grazing cattle seemed each a moving
 dot.
Faint from the bustling village came murmurs
 low and deep;
The bells far off did tinkle ; the lad lay fast asleep.
Asleep he lay, but not for long—he woke; a
 grimy hand
Pressed his mouth! His wrists were bound!
 Around him Gomez's band !
They dragged him to the convent; cried Gomez,
 " Rope this fool !"
Then 'neath the rope they placed him, kneeling
 upon a stool.
Around his neck so slender the snaky bell-rope's
 fold
They fastened. Then cried Gomez, " That bell
 won't soon be tolled !

Come on, lads, there's work below ; this fool ain't
    to be hung,
By the saints ! yet hang he will before that bell
    is rung !"
The robbers laughed and vanished, and Rooe was
    left alone
With one thought ever stinging—he must his
    fault atone.
The rope his throat was galling, his corded
    wrists were numb,
Poor Rooe's burning thoughts alone could freely
    go and come :
The helpless souls, the bell above, the black band
    creeping down.
Over his brow the drops rolled fast—he must
    arouse the town !
That rope, he well remembered, his strength had
    often tried,
And all his weight to move it he knew must be
    applied.
He thought of home and mother, of Carmen,
    sweet and fair,
Then, with one sob of anguish, he sprang into
    the air !

    .   .   .   .   .   .

The robber band was creeping down the steep
    incline
With Chieftain Gomez leading the dark, exulting
    line.
"They're ours," the bandit chuckled, "it's time
    to make the charge,"
And then the robbers halted upon the hill-top's
    marge.

Red Gomez drew his sabre, and then—What was
    that sound?
Bom! Bom! The convent tocsin! It fairly shook
    the ground.
Bom! Bom! Pale grew the robbers, yet Gomez
    cried, "Advance!"
Too late, the town was rousing, and lost the ban-
    dit's chance.
Some scattering shots! The robbers fled over
    the hill-top's crown.
Bom! tolled the bell yet fainter—saved was the
    little town.
Straight upward strode the peasants, up to the
    convent tower,
Before them sways a something—from which the
    bravest cower;
Bom! clanged the bell yet fainter, and with the
    passing toll
Its dying sob bore upward the hapless Rooe's
    soul.
They took him down with wailing, and bitter
    tears were shed,
For he who saved Zanora, mute as its bell—was
    dead.

                                    W. R. Rose.

www.ingramcontent.com/pod-product-compliance
Lightning Source LLC
Chambersburg PA
CBHW021356210326
41599CB00011B/898